FOCUS ON

HITCHCOCK

❖❖

edited by

ALBERT J. LaVALLEY

A SPECTRUM BOOK

Prentice-Hall, Inc.
Englewood Cliffs, N.J.

FILM FOCUS

Ronald Gottesman and Harry M. Geduld

General Editors

THE FILM FOCUS SERIES PRESENTS THE BEST THAT HAS BEEN WRITTEN ABOUT THE ART OF FILM AND THE MEN WHO CREATED IT. COMBINING CRITICISM WITH HISTORY, BIOGRAPHY, AND ANALYSIS OF TECHNIQUE, THE VOLUMES IN THE SERIES EXPLORE THE MANY DIMENSIONS OF THE FILM MEDIUM AND ITS IMPACT ON MODERN SOCIETY.

ALBERT J. LAVALLEY *is Associate Professor of English at Livingston College, Rutgers University, where he devised a film studies program and teaches history of film. He is the author of* Carlyle and the Idea of the Modern *and editor of* Twentieth Century Interpretations of Tess of the D'Urbervilles.

© 1972 by PRENTICE-HALL, INC.
Englewood Cliffs, New Jersey

A SPECTRUM BOOK

ISBN: C   0–13–392373–8
P   0–13–392365–7

Library of Congress Catalog Card Number 78–163858

Printed in the United States of America

10   9   8   7   6   5   4

PRENTICE-HALL INTERNATIONAL, INC. (*London*)
PRENTICE-HALL OF AUSTRALIA, PTY. LTD. (*Sydney*)
PRENTICE-HALL OF CANDADA, LTD. (*Toronto*)
PRENTICE-HALL OF INDIA PRIVATE LIMITED (*New Delhi*)
PRENTICE-HALL OF JAPAN, INC. (*Tokyo*)

CONTENTS

THE FILMS

ACKNOWLEDGMENTS

In preparing this book, I was guided, inspired, and challenged by my friends Ronald Gottesman and Leo Braudy. Their suggestions and sharing of ideas played a major role in shaping this project. I am deeply indebted to them. Along with my colleague David Leverenz, they also carefully read the introduction and made valuable suggestions for revision. My thanks for this, too.

Many other colleagues and friends also helped. My thanks in particular to George Levine, Gerald Rabkin, Ronald Christ, Leslie Clark, Roy Skodnick, Miguel Algarin, Susan Braudy, Robert Fraker, Stuart and Leslie Mitchner, Alan Schiffmann, and Wheeler Dixon. Richard Henshaw of Cinemabilia shared his vast knowledge of writings on Hitchcock and located many hard-to-find articles. Ian Cameron and Raymond Durgnat generously offered information about recondite writings. Several authors kindly offered articles for which there was finally, alas, no room—among them Ernest Havemann, Art Buchwald, Robin Gadjusek, and Mr. Hitchcock himself. My thanks to all of them.

In researching the book, I am grateful to the staff of the theater collection of the Library of the Performing Arts at Lincoln Center in New York and to that of The Museum of Modern Art's Film Study Department, particularly Mary Corliss. Peter Nevraumont of Universal-Kinetic and Beatrice Hermann of Audio/Brandon made it possible for me to see several Hitchcock films.

My film students were rich with suggestions when we studied Hitchcock. Jim Krell, Eric Krueger, and Steve Daley were inspiring in many ways. Joanne Rynk and Justine Trueger worked long hours on the line drawings for the *North by Northwest* analysis. I would also like to thank Amy Snyder, Anne Riedel, Raisa Terebey, George Spozarsky, Henry Sakow, Tracy Shagin, and my two special students, Val Gottesman and Marge Levine. The translations were done by Stephen Arkin, Alan Schiffmann, and Jean-Louis Guillois. My thanks to them, too.

In the editing of the volume, I was fortunate to have Claudia Wilson of Prentice-Hall working with me. The book presented unusual problems and went through many complicated changes; she was always ready with clear solutions and good suggestions about the material. I am enormously indebted to her. Harry Geduld, the co-editor of the Film Focus series, also made good suggestions about the book's format.

Finally, I want to express a general indebtedness to my late parents, Fred and Lillian LaValley, who encouraged my excitement about film when I was a child and patiently listened to my endless recounting of Hitchcock movies. To them this book is gratefully dedicated.

Introduction
by ALBERT J. LaVALLEY

Alfred Hitchcock is certainly the only great film director whose name, figure, manner, and general thematic concerns are well known to the public. Largely through his regular and always bizarre television appearances as host of *Alfred Hitchcock Presents* and his walk-ons in his own movies, he has become familiar, in body and spirit, to millions. His peculiar blend of comedy and melodrama has raised such expectations about his films that his name is frequently coupled with his stars on the marquees of theaters—rightly so, since he is one of few directors whose name can draw the general public into the theater. For a long time he has been "the master of suspense," an *auteur,* known for his cinematic style long before the word gained currency in film critics' circles.

Unfortunately, Hitchcock has sometimes seemed to be imprisoned in the role. The title "master of suspense," or even master of film technique, carries with it a reverse snobbery. If he is an *auteur,* isn't he a *petit auteur* limited to mere melodramatic genres? Haven't higher themes remained beyond his reach? Doesn't he just provide fun and games, a ride on a roller coaster, to which he has been known occasionally to liken his own films? Doesn't he dilute his occasional serious suggestions with comic bits and gags, thereby deflating their impact? Isn't his irony merely cute or evasive? To many serious film critics Hitchcock has been a disappointing figure, a great visual artist with a penchant for "pure" cinema, a master of montage and narrative technique, who wasted his gifts on melodramatic genres. Some critics will add patronizingly that he did those gestures well, especially during his English period. Occasionally, perhaps, in films like *Vertigo* (1958) and *Psycho* (1960), he rose beyond his limitations.

By contrast, other critics, particularly the young French critics of *Cahiers du Cinéma* in the 1950s (Truffaut, Rohmer, Chabrol, Astruc, Domarchi—many to become film-makers themselves) assumed that Hitchcock could do no wrong. They reversed the popular notion that Hitchcock's English suspense movies of the 1930s represented the perfection of his limited art and the American movies a dilution of his skills. Instead, they saw in the American Hitchcock a serious explorer of metaphysical anxieties whose cinematic methods and themes progressively deepened. While the English critics bemoaned the disappearance of his crisp pacing and cutting, noted the absence of the usual sharp glimpses of social realism, and blamed Selznick, Hollywood, and the influence of the star system for both, the French celebrated the way the films opened out into murkier psychological areas, their subjective quality, and the long, mysterious, and troubling tracking shots that were added to the methods of Eisensteinian montage and quick cutting. Beyond this the French even claimed a "noble" thematic consistency to Hitchcock's work, allying it with Dostoyevsky, Kafka, and Poe, and they buttressed their moral argument with appeals to Hitchcock's Catholic upbringing and Jesuit schooling. Indeed, when Eric Rohmer and Claude Chabrol wrote the first book on Hitchcock in 1957 (since then there have been about nine others[1]), they claimed to have discovered a structural pattern that was the key to Hitchcock's films: a transfer of guilt in which an innocent weak figure yields to a stronger evil figure and thereby in some way shares the evil desires of the latter. For Rohmer and Chabrol Hitchcock's movies confirm the Catholic doctrine of original sin, that all of us, no matter how innocent we may think we are, are guilty. Whatever its value—some would claim it is the "original sin" of Hitchcock criticism—the Rohmer-Chabrol thesis had the effect of launching a major debate, which has been conducted ever since under the shadow of its terms.

In the midst of excessive detraction and praise, of interpretations that are caustically negative or elaborately solemn, Hitchcock has remained his detached, modest, and frequently ironic self. A master of the art of being interviewed, Hitchcock has been remarkably consistent over the years in his attitudes towards film, the mechanics of suspense, what he expects film to do, how he employs the montage method, and the genre of melodrama. He has good answers to all questions, but he resolutely refuses to be led into metaphysical or high moral claims. A tireless publicist of his own films, he usually says much the same thing about each one, invoking the same principles of suspense and merely changing the name of the film under discussion. His refusal to be led to see what his French admirers have discovered has exasperated them, providing fuel to those who want to undercut suggestions of a serious import in his films. André Bazin, the dean of the *Cahiers*

[1] See this volume pp. 180–81 for a listing of other books on Hitchcock.

critics and a "cautious" admirer, recounts in "Hitchcock versus Hitchcock" [2] how he convinced Hitchcock about a pattern of identification of weak with strong which permeates his work—a transfer of role and, by implication, of guilt. Anyone who reads his tale of this process will realize it was not an easy one, nor was its resolution completely satisfactory. When François Truffaut proposed that "most of your work is strongly permeated by the concept of original sin, and of man's guilt," Hitchcock snapped back, "How can you say a thing like that when in fact we always have the theme of the innocent man who is constantly in danger, although he isn't guilty?"

Truffaut then tries to clarify matters:

> While your hero is generally innocent of the crime for which he's under suspicion, he is generally guilty of intentions before the fact. For instance, let us take the character of James Stewart in *Rear Window*. Curiosity isn't merely a nasty personal trait; in the eyes of the Church it's actually a sin.

Hitchcock acquiesces, but shifts ground away from morality to aesthetics:

> That's true, and I agree with you. You remember that a reviewer said it was a horrible film because of the Peeping Tom character. Now, if anyone had mentioned that to me before I embarked on that picture, it certainly wouldn't have kept me from going ahead with it, because my love of film is far more important to me than any considerations of morality. [3]

The effect here is a bit vertiginous. Hitchcock seems to refuse to be pinned down. Is he pirouetting, indulging in some defense mechanism, donning some "intellectual camouflage," as André Bazin suspects? Or is he merely evading Truffaut's insistence on some will-o'-the-wisp overall theory? And even if he is, to what extent is Hitchcock conscious of doing so? He remains an enigma. Bazin is right to suggest that creation may be intuitive, not merely rational and self-conscious. Better to turn to the films themselves, perhaps even to Hitchcock's preplanning techniques, where he sketches out each scene in advance and conceives the narrative visually in his head before putting it on film. Here Bazin can reach

[2] André Bazin, "Hitchcock contre Hitchcock," *Cahiers du Cinéma*, no. 39 (October 1954). This article also appears in *Cahiers du Cinéma in English*, no. 2 (1966) and in this volume, pp. 60–69.

[3] François Truffaut, *Hitchcock* (New York: Simon & Schuster, 1967), p. 240.

a Hitchcock that is both serious and suspenseful. In any shot, even at this early stage, Bazin notices:

> a tension that one would not know how to reduce either to dramatic categories or plastic categories but which partakes of both at the same time . . . creating in the *mise en scène* . . . an essential instability of image. Each shot is thus, for him, like a menace or an anxious waiting.[4]

Nevertheless, certain aspects of Hitchcock's talk—usually unexamined offhand remarks—clearly point to serious themes. His often-told account of the origins of his fear of the police (he was sent by his father at the age of five to the police with a note, only to discover that it told the policeman to lock him up as a punishment for being naughty) has about it a traumatic note, as though it has been translated into ritual autobiography, much like Dickens' account of his childhood days in the blacking house. Hitchcock's testimony about his Jesuit education contains no such repeated story, though he clearly stresses the authoritarianism of his teachers and the fear it produced in him. Similarly, Hitchcock's association with Ufa (Universum-Film-Altiengesellschaft) in Germany and the influence of Lang, Murnau, and German expressionism influenced him thematically as well as stylistically, but he shows no inclination to talk about it.

In the matter of aesthetic theory, Hitchcock has always been interested in the idea of the pure film, even upon occasion wanting to make plotless films about travel or the passing of a single day (a favorite subject with the *avant garde* filmmakers of the 1920s). "Pure cinema is complementary pieces of film put together, like notes of music make a melody."[5] Elsewhere he describes it as "pieces of film put together . . . artfully, and creating ideas."[6] Critics have always noted this quality in his films, but for many it remains mere formal excellence, an aesthetic delight that goes nowhere. "One can, of course," notes Raymond Durgnat, "go to every Hitchcock movie confident of enjoying a smooth, confident drive, as in a Rolls-Royce of style, along a complex run of moral and dramatic landscaping, whose expertise is in itself a legitimate source of purely aesthetic pleasure."[7]

[4] Bazin, p. 32.

[5] Quoted in Peter Bogdanovich, *The Cinema of Alfred Hitchcock* (New York: Museum of Modern Art, 1962), p. 4.

[6] "Interview with Alfred Hitchcock," *Take One* 1, no. 1 (September–October 1967): 17. See this volume p. 22–27.

[7] Raymond Durgnat, "The Strange Case of Alfred Hitchcock," Part II, *Films and Filming*, April 1970. The series of articles ran from March to September 1970. Part III is reprinted in this volume, pp. 91–96.

Yet the analogy with music that Hitchcock repeatedly invokes does not suggest the neglect of theme and plot so much as the purification of experience, a kind of emotional rhythm, which the camera, like an orchestra, will follow. Elsewhere Hitchcock amplifies this:

> Now you see when I'm on the set, I'm not on the set. If I'm looking at acting or looking at a scene—the way it's played, or where they are—I am looking at a screen, I am not confused by the set and the movement of the people across the set. In other words, I do not follow the geography of a set, I follow the geography of the screen. I can only think of the screen. . . . I say (to the cameraman photographing an actor going to a door) "Well, if he's still in a mood—whatever mood he's in—take him across in a close-up, but keep the mood on the screen." [8]

Hitchcock is fundamentally interested in the narrative line, not in a message, not even in the significance of plot or character, but in the development of emotional resonances in narrative through cinematic method. Of *Psycho* he tells Truffaut:

> People will say, "It was a terrible film to make. The subject was horrible, the people were small, there were no characters in it." I know all of this, but I also know that the construction of the story and the way in which it was told caused audiences all over the world to react and become emotional.[9]

There is a slight air here of evading thematic concerns, but at the same time there is the conviction that the movie is more than its themes, that it is an emotional curve Hitchcock shares with—and practices upon—his audience, that digs beneath their complacencies and everyday life into something deeper, which resists any total reduction to character, plot, theme, or intellectual structures. In short, Hitchcock acknowledges the perceptual power—the magical and mythical dimensions—of the film experience.

Undoubtedly the structures of doubling and shared guilt that the French critics have made much of are also present in the movies. But they too have a musical analogy, a kind of formal scaffolding that permeates and enriches the emotional line, but never defines it adequately. Movies that the French are attracted to—*Shadow of a Doubt* (1943) (the guilty uncle and the innocent niece are both called Charlie; camera shots of one echo shots of

---

[8] Quoted in Bogdanovich, p. 4.
[9] Truffaut, p. 211.

the other), *Strangers on a Train* (1951) (Bruno and Guy, opposite but complementary, "exchange" murders), *I Confess* (1952) (a priest hears the confession of a man who was blackmailing him), and *The Wrong Man* (1957) (an innocent man is accused of a look-alike's robberies and his wife goes insane during the ordeal of proving his innocence)—all have as well a dizzying internal set of doubles that echo the principal ones. These repetitions finally have a vertiginous and nihilistic effect on us instead of establishing rigorous moral and intellectual patterns. The patterns above do not account for the different feeling, texture, and resonance of each of the films. Furthermore, when they are most insisted upon, as they are in the latter two films, which also have theological structures, they seem to fail and the movies lose their power. They become failures for Hitchcock too. The real strength of the films lies elsewhere.

Behind the rather clever and unconvincing façade of the two Charlies in *Shadow of a Doubt* (1943) lurks the sense of a vacuum shared by the two figures, the world of an aimless, unexciting small town waiting to be filled with the knowledge of emptiness and evil that Joseph Cotten brings. Cotten becomes a vehicle for exploring in some depth the loneliness and isolation that are already in the young niece Charlie. The film drives towards a disquieting knowledge of loneliness, of something incommunicable in each person, an experience of unbridgeable gaps at the heart of individual life —despite the similarity of names and relationships. *Strangers on a Train* (1951) uses its doubling not to promote the idea of doubling and universal guilt but to foster various kinds of psychological unease. Indeed, almost anything can be a double of anything else, so all appearances are deceptive and threatening. Beyond the "exchange" of murders, there is the undercurrent of sex and murder, of homosexual threats and attractions, the blending of wives and mothers, parental revenge and marital fights, even the ghostly suggestion of one person reincarnating another. Hitchcock is not organizing our experience into structural systems, but rather using structures to release a kind of absurdist logic in life. ("Logic is dull," says Hitchcock. "The fact is I practice absurdity quite religiously." [10]) He makes life seem dreamlike, its surface a thin crust over a substratum of fear, insecurity, unconscious anxiety, and guilt. In this dream world one character evokes another, one experience pulls together threads of many past ones. Hitchcock is indeed interested in guilt and innocence, but not in a metaphysical or intellectual way; he wants to know what it feels like to be guilty, to have handcuffs on, to undergo anxiety. All his camera techniques are concentrated on rendering this emotional texture. Furthermore,

[10] Ibid., p. 194.

he is interested in linking these states up with other states of the unconscious. Raymond Durgnat puts his finger on this vision when he suggests:

> The real Hitchcock touch is a far more diffuse affair than a moral schema or points of style. . . . There is a sense of having penetrated from an apparently tolerant, even permissive, world to a grimmer one, whose cruelty seems, confusingly, both amoral and morally unremitting. In the British films an everyday world of familiar foibles and eccentricities parts to reveal grimmer patterns. . . . But in all cases the richness of experience lies not so much in the vision from which one penetrates, so much as the process of penetration. Hitchcock's films can't be justified by reference to any one "layer": Their artistic impact is in the intermeshing of layers.[11]

By these criteria, Hitchcock's American films, more diffuse and more psychologically subjective, joining more genres together, do show a greater richness, even if during the 1940s they were often less successful. As John Russell Taylor suggests, we can regard the great thirties films as "the perfection of a style," and still see the forties as "the appreciation of the limitations of that style and an erratic quest for a new style" and the fifties and sixties as "final maturity." [12] The notion of American versus British Hitchcock has been badly overstated and has encountered stiff resistance and there are signs that it is breaking down. Even Penelope Houston, who cannot find "a figure in the carpet," admires *Vertigo* (1958); and Raymond Durgnat, no champion of Hitchcock in general, clearly thinks *Vertigo, Rear Window* (1954), and *Psycho* (1960) are great films.

A key to this intermeshing of themes is Hitchcock's continuing interest in romance, from the early silent pictures on. Robin Wood regards the winning of the woman, which concludes many of the films, as a reward for the hero, who has successively matured through the therapy of the anxious experience the movie gives us. My own inclination is to see things in a murkier light. Hitchcock's concept of romance includes fetishism, sex, and murder. These are frequent topics of Hitchcock's conversation and usually the most startling aspects of it—though again they have been little examined. For instance, with Ian Cameron and V. F. Perkins he had this exchange:

---

[11] Durgnat, Part II.
[12] John Russell Taylor, "Hitchcock," *Cinema Eye, Cinema Ear* (New York: Hill & Wang, 1964), p. 171.

C. and *P.* In *North by Northwest* Grant seems to want Eva Marie Saint dead; he's happier when she seems an enemy or in danger than when she seems to be an available wife or lover.

H. What's that old Oscar Wilde thing? "Each man kills the thing he loves." That I think is a very natural phenomenon, really.

C. and *P.* You don't find it somewhat perverted?

H. Well, everything's perverted in a different way, isn't it? [13]

And with Peter Bogdanovich:

B. Wouldn't Kelly [in *To Catch a Thief*] prefer Grant were really guilty of the robberies?

H. Oh, of course. Let's put a mild word to it—it's more piquant that way, more in the nature of her fetish.[14]

Also to Bogdanovich, he describes James Stewart's remodelling of Kim Novak in *Vertigo* as an undressing of her.

Sex is a central experience where feelings of other kinds— anxiety, intensity, fear, and the unexpected particularly—abound. There is a desire on Hitchcock's part to link up with orthodox Freudianism as the system that most closely approximates his vision of the unconscious, but the films never quite connect. It is not just that the Freudianism of *Spellbound* (1945) is naïve, but that it really has little to do with the major intensities of the movie. Scratched parallel lines on a bedspread, the mixture of sexual and murderous attitudes, are more rich than Ingrid Bergman's solution of the murder via psychoanalysis at the film's end. *Marnie* (1964) coalesces two major themes of Hitchcock films, the dominating mother and sexual neurosis, but again the disquieting effect of the film comes from the odd interaction of characters, the strangeness of speech and motive, a perverse experimentation with Tippi Hedren by Sean Connery, and the obsessive concentration of the tracking shots of Tippi Hedren; it has little to do with Mom at the end. The movie is more troubling than that final scene suggests. The same is true of the psychiatrist at the end of *Psycho,* whom Hitchcock seems eager to parody and dismiss. Indeed he does so with the powerful final images of Perkins against a white wall, mother's voice and skull, and Janet Leigh's car rising from the murk—the murk of the mind.

Even in the English films the mixture of sex and mystery is

[13] Interview of Hitchcock by Ian Cameron and V. F. Perkins, *Movie* 6, January 1963.
[14] Bogdanovich, p. 33.

troubling. The overlap adds to the disturbing quality of the whole. *Blackmail* (1929), for instance, contains a scene in which Anny Ondra is seduced by Cyril Ritchard. Lindsay Anderson finds it badly played. But in fact the disturbing quality of the scene persists even if one accepts Anderson's condemnation of the acting. Anny Ondra is coy and cute, seemingly innocent, but really eager for sex; furthermore, she has just fought with her boyfriend, a detective, and her motives for picking a new man up are disquieting. Outside his house there is another man; Ritchard is stopped in the hallway and given a letter. So when the seduction scene takes place, there is an uncertainty to the experience because many possibilities seem ready to emerge. We don't know if he will murder her (he seems more threatening than interested in sex), or if she will kill him if he attacks her, or if both of them will be threatened by something outside the house. We are further troubled by the sudden movement from expectations set up by the genre of social comedy to uncertainties released by their violation. We are caught in an emotional state where characters do not count, but the release of the emotion in almost any direction does. We are then both satisfied and horrified that Anny Ondra murders Ritchard for trying to seduce her—more so now in retrospect when the morality of the situation has changed somewhat.

The endings of *Blackmail* and *Sabotage* (1936) are cases where the troubling note is sustained by maintaining the sexual motif. Hitchcock claims that he wanted to end *Blackmail* ironically with a restatement of the beginning, this time with the detective arresting his girl for the murder (an example of Hitchcock's fascination with formal perfection as well as irony), and he claims that the producer's box-office consideration forced a compromise, hence the romantic ending. But the ending is scarcely romantic. True, the detective goes off with the girl. But the clown portrait at the scene of the knifing sardonically laughs at her as it is carried down the police corridor, and though the girl has confessed to the cop, he has protected her and both are now guilty. Romance coexists with a kind of psychological malaise of mutual guilt; they are the real "blackmailers" of the movie. *Sabotage* also lets its female protagonist get away with the murder; Sylvia Sidney confesses to the police, but they do not hear her, so preoccupied are they with the explosion of the movie house. Her husband's body—the evidence that would convict her—disappears as a threat and the film allows the detective to take her away; again the guilt seems to diffuse itself over both characters, the whole film—and us.

In the more comic films, the movement from sex to mystery is more alternating. *The Thirty-Nine Steps* (1935) has a more buoyant tone; we concentrate less on all the images of bars, searchlights, shadows, and handcuffs that embrace Robert Donat than on his adroitness in getting out of terrible situations. Still the movie is filled with varieties of sex that are disquieting, however comically rendered. Donat is handcuffed to a woman who thinks him guilty and hates him, while he is gradually attracted to her. The unhappy married couple, as Pauline Kael notes, are handcuffed without handcuffs. Even the murder of the agent, the famous scene of the woman slumped over Donat's bed with a knife in her back, suggests a sexual scene, and Donat's excuse in leaving the house—when he finds the truth won't work—is a story of sexual indiscretion which he tells to a milkman. Later in his train compartment, two eccentric salesmen, reading the paper in front of Donat, speculate on the crime as a sex murder and alternate their own talk of murder with discussions of female underwear. Hitchcock has said that the Donat character should have been more careful in protecting the woman and that he operates out of guilt. The love affair analogy can hardly be missed.

But it is in the American films where this mixture becomes most disquieting. If one measures *Notorious* (1946) by the pace, characterizations, and methods of the British films, it is a failure. But seen as a subjective adventure into morbid eroticism and primitive fears, the picture is quite successful.

In the service of this new kind of adventure, Hitchcock has grafted several genres together and let them—and the expectations they set up in us—interact. There is the spy story: no matter what happens we want Grant and Bergman to get the secret out of Claude Rains' house. There is the romance: no matter what happens we want Grant and Bergman to get each other. Yet the two worlds, public and private, cannot be kept separate; the stories—and our feelings—blend. The confusion is rendered even more unsettling because of the third story and the interest it awakens in us. Rains is the villain, but he is treated sympathetically. He really loves Bergman and in marrying her he tries to free himself from his mother's domination. There is the suggestion that his mother is the real Nazi, the true villain, and that Rains' decency, love, and trust will be the source of his purification. Consequently, we want him to be right too. By marrying her old friend Rains, Bergman does what we want her both to do and not to do. She is both ennobled and sullied. That Hitchcock does not explore her moral dilemma in great depth is not, it seems to me, a serious objection to the film, as some critics have alleged. Hitchcock's interest is in the general undercurrent of fear and tension, the confusion of emotional and moral states, that the narrative line of all three stories evokes in us.

Hitchcock's camera centers subjectively on Ingrid Bergman, for she is the one who has to undergo the most anxiety, posing as Rains' beloved. But the camera also lingers subjectively over the faces of Rains and Grant. Rains becomes more sympathetic than our generic expectations usually allow because he truly loves her and because he is being practiced upon; Grant becomes less sympathetic because he suffers in silence. He is inactive and forces Bergman to go into the house. Significantly, we frequently see only the back of his head—an impersonal and cold rendering of him. From the beginning he is attached to a chilly bureaucracy of the public world, represented by Louis Calhern bedecking Bergman for seduction by Rains or callously eating cheese and crackers in bed while he discourses distantly about danger. Though Hitchcock in World War II treats the Nazis as villains and the American cause as right—and later nominally treats the Cold War Russians in *Topaz* (1969), *Torn Curtain* (1966), and *North by Northwest* (1959) in the same way—he manages to rise above flag-waving patriotism, or in the case of the Cold War narrow American advocacy, by subsuming his heroes as bureaucrats into a larger theme of human loss and stultification.

This confusion of genres and sympathies for characters evokes a certain "primitivism" in the plot. Our own insecurities, our movement towards a state of childish passivity, are reflected in the way the plot takes on overtones of a fairy tale and in its archetypal collision of forces. We want "Prince Charming" Grant to rescue "Princess" Ingrid Bergman from "the ogre and his castle." We want "father-figures" to be defeated. Significantly, Rains is never really seen in any sexual encounter with Bergman and since he is older and a friend of her father's, we are induced to take him as a father rather than a lover. But he too participates in an archetypal story. Hitchcock usually reserves his real terror for mother-figures; it is Rains' mother who is the real power, the "witch of the castle." Rains is noticeably much shorter than Bergman and in his mother's room he appears a miniature figure huddled up in a chair, while she appears as a strong dominating image. When she toughly lights the cigarette, upon learning of her son's deception by Bergman, her fears confirmed and her domination regained, she is truly masculine.

The disquieting emotions the film produces are closely linked to its methods as well as to the weird blend of sympathies for the three major characters. The deep-focus camera work and the tracking methods reinforce this primitivism at all points. Like a child, we take it all in, watch and wonder, are curious and on guard. Because of the lying and deception involved in the plot that Bergman undertakes, we are doubly tense in the house, reduced to caution, but also to childlike fears, which the tracking camera reveals. The house is awesomely impressive, intimidating, mysterious when we see it through Bergman's eyes the first time. The appearance of Rains' mother, coming from long shot, slowly gliding down the stairs, to intimate close-up with the camera as Bergman's eyes, is terrifying. The witch of the palace converges upon the victim. The shot is held for a long time; we must wait for her to come down the stairs. We feel the anxiety, the waiting, the getting ready for the deception. In this house, the good people as well as the bad must lie. When Rains' mother stops at the camera, which functions as Bergman's and our eyes for the moment, we are suddenly accused and on guard, made to feel the discomfort of our deception, our fear. Similarly the camera lingers just a moment too long over each of the introductions of the spies to Bergman; we are conscious of her double attempt to be ingratiating and to remember their names. But we are also conscious of something sinister and frightening in them. (A characteristic Hitchcockian reversal is to make the scientist she informs on the most sympathetic of the lot.)

Silences in the film emanate terror, suspicion, distrust; the world is not really as it seems. And we as an audience do not quite know where we stand, morally or emotionally. The great stairway of the house, over which Hitchcock's camera glides, becomes a great obstacle, an avenue to the prison of Bergman's room, a massive architectural intimidation. The dramas played on it are less the lofty dramas of power or romance usually suggested by such staircases than they are evocations of primitive, almost childlike insecurities that stairways, shut doors, imprisoning rooms, and father figures can evoke. The slow poisoning of Bergman reduces her to the helplessness of a child who cannot manage the stairway, who has to be put to bed and shut in her room. Hitchcock's tracking shots make us feel that serious threats reside in the simplest objects; all things are insecure, seen as though from the perspective of a child who has not learned how to orient himself. In a basically insecure and threatening world loyalties are ambiguous and parent figures are more threatening than protective. Both here and in *Suspicion* (1941) Hitchcock evokes the primitive fear of drinking poison that probably underlies any act of drinking we may do when someone else has handed us the glass (significantly, in *Suspicion* the drink is the "bedtime" glass of milk); at any rate, he suggests how long and slow a process it is to build up trust—and how swiftly it can be compromised, especially by public loyalties, whether Rains' Nazism or Calhern's patriotism.

During the 1950s and 1960s Hitchcock achieved full mastery of the methods and themes suggested by *Notorious* (1946). Most of the articles on particular films in this anthology are concerned with these films: *Strangers on a Train* (1951), *North by Northwest* (1959), *Rear Window* (1954), *Vertigo* (1958), *The Wrong Man* (1957), and *Psycho* (1960). The critical articles clearly show the enormous dimensions and range of Hitchcock's art. For me they confirm Truffaut's thesis that the American films, particularly those of the fifties, offer a rich broadening out of themes and methods, a maturing of vision and craft. Let me add that they also offer new complications with their growing self-consciousness, their overt treatment of voyeurism, both as subject and method, and their self-parody.

In his later films of the 1960s Hitchcock goes in still new directions and often deliberately breaks his usual methods of suspense, defuses what suspense has been building up, to pull us into a more reflective mood. He draws us away from what is happening to make us think about what it implies, about the relationship of one event to another, of event to character, of the expectations of genres to reality. He ponders his patterns. In these later Hitchcock films—what we might call the films of his old age—the usual direct and frightening experiential quality of the other films is muted; the scaffolding, the forms, the patterns that played around all his earlier films take center stage. They are not seen as definitive or fixed; their inherent absurdity is underlined. The quietly ironic mood of a detached yet puzzled artist dominates; there is an openness to these films because the aesthetic patterns are no longer asked to round things out (though they never really did anyway). All of them have a certain surreal and nightmarish quality; the forms of daytime—the way we organize our lives, our emotions, our social reality—are shown as strange, discordant, and arbitrary. The heroic scientist of *Torn Curtain* (1966) keeps turning into a comic figure, unable to dominate the illogic of events. In *Topaz* (1969), the movements of various governments and their interests push the characters like robots through the action and into tragedy; the arbitrariness of unlikely places and people colliding takes precedence in its effects over any of the protagonists and their problems. Each of these four films (*The Birds* [1963] and *Marnie* [1964] are the remaining two) has about it the air of a final statement in a major area of concern for Hitchcock during his career: *The Birds* on the relations of civilization to nature and our own basic nature; *Marnie* on Freudian fears and anxieties, *Torn Curtain* and *Topaz* on the cruel disjunctions of public and private worlds, the ultimate impersonality of the public world and its priorities. There is even an element of allegory and a sense of a cosmic canvas in *The Birds* and *Topaz*.

Of the four, *The Birds* seems to me to be the most successful, though the interplay between its realistic intent and its more stylized allegorical methods is not fully realized. (I am not quite sure how we are to relate to the characters. Are they real people? Ciphers as real people? Or abstract statements? There are gestures towards realism in portraying them, but I am frequently bothered by the half-heartedness of such gestures—the awkward acting, the bad process shots and back projection, the unreal attacks of the birds. Some will argue that such gestures recall us to the wider implications of stylism, to allegory and broader meaning. Unfortunately I get locked into the bad attempt to mediate between the two.) Whatever its difficulties, the ambitiousness of *The Birds* is evident and often, especially in its drive towards allegory, quite successfully realized. A meditative movie, it uncovers the spaces between people as well as the façades of civilization. Hitchcock notes various resemblances between birds and people, especially in the stylized acting of Tippi Hedren. The force of what at first seems coy, playful social comedy fully emerges when the behavior of the birds becomes arbitrary, when they turn from peaceful creatures into nightmarish killers. Having established his parallels, then more strongly grounding both birds and people in nature itself, Hitchcock is able to suggest the insecurity behind all relationships: mother and son, husband and wife, lover and beloved, the people in a town. The behavior of the birds is the final rough truth of nature that smashes through the cartoonlike quality of real life.

But Hitchcock does not want his characters in *The Birds* to become complete cartoons. He wants them to be real people too and he concentrates on their individual problems in great detail. He sketches his people with some depth of individual response. Carefully they fight against being pulled into the night world of chaos and disintegration, carefully they rebuild their trust, even though over the final shot the atmosphere of insecurity provided by waiting birds prevails, even though nature—human nature too—cruel and unremitting, is somehow ultimate. In refusing to yield to a total cartoon, a science-fiction or horror movie, Hitchcock suggests his apocalyptic vision and yet counters it; his realism and concern for people suggest the sources for an optimism that has allowed him to make comedies and romances and infuses the elements of these genres in all his films.

Hitchcock's movies have always portrayed a world dissolving into arbitrary chaos and he has delighted in producing a kind of emotional chaos in his viewers, perhaps never more so than in *Psycho* (1960) where he destroys in rapid succession our three identification figures. Some have seen in this Hitchcock's perverse pleasure in playing God and toying with us. According to this theory, he does not share these fantasies but merely manipulates us with them and scorns and hates his audience. But Hitchcock's detachment need not be either a lack of concern or a diabolical manipulation. His themes and interests are indeed almost too obsessively recurrent not to be his own. His montage methods, emphasizing shock and outrage, tend to portray the world as topsy-turvy; his tracking shots show it as threatening and filled with magical, fearful objects. That he is in control of producing effects he has obviously experienced shouldn't be disturbing; his methods are finally defenses against submitting to a world of chaos, means of attaining a perspective on it, while acknowledging its power.

Hitchcock's viewpoint seems to me close to nihilism, though he would call himself a strong moralist. Perhaps finally he is not in control of the dark forces of the mind which he releases and lets us experience. No one is. The forces simply exist. But against them he suggests the painfully built-up aids of mutual trust and ironic detachment. His films finally have a surreal thrust to them. Characters and action are like a kind of dream, not misty and foggy as in a work by a German expressionist, but sharply outlined, yet opaque, mysterious, acting in unlikely ways. Such precise, cool, and vivid images are not at all like the traditional images we have of dreams, but are probably more like actual dream activity itself.

Though Truffaut sees Hitchcock not as a fantasist but as a realist, in truth he is both. His films never surrender their hold on real life and the social world, but at the same time they tap the inner life of his characters and his audience. The great brooding statuary heads in the British Museum in *Blackmail* (1929) and of Mt. Rushmore in *North by Northwest* (1959) show man as a diminished creature, running in flight and fear over a threatening cosmos. But they also imply the godlike and rather inscrutable and ironic detachment of a creative stance that can let the dreams happen, an attitude that allows him to regard the horrors of *Psycho* as therapeutic fun, a joke, valuable for life itself.

Part 1 of this book allows Hitchcock to speak for himself in interviews and articles. The interview from *Take One* is typical of Hitchcock's public relations with the press and young film critics. Casual, relaxed, witty, and voluble, he sets forth his basic principles of film technique (montage and preplanning) which he has enunciated from his earliest days of making pictures. Only in the interviews with Peter Bogdanovich about specific films does he reveal some of the moral ambiguities and complexities that the *Cahiers* critics celebrate. The two articles, "Direction" (1937) and "Rear Window" (1968) offer the reader a chance to see continuities and contrasts between the English Hitchcock of the 1930s and the American Hitchcock of the 1950s and 1960s. In the second, Hitchcock has settled into his comfortable public role; he tells some of his favorite stories and talks about some more basic principles of film-making. But in the first, one can feel him still discovering his craft.

Parts 2 and 3 offer a wide range of Hitchcock criticism, both pro and con, ranging from the journalistic to the speculative and metaphysical. I have, however, omitted the strongly negative or merely vituperative criticism—of which there is a fair amount—in the belief that criticism that has some positive feeling for its subject is of more value.

Part 2, "Hitchcock Controversy," affords the reader some sense of the developments in Hitchcock criticism and its central issues. Lindsay Anderson's 1949 assessment of Hitchcock's career, with its praise of the English films and its scepticism about the American films, represents the dominant attitude of the time—and is still widely held by many responsible critics, among them Raymond Durgnat, Penelope Houston, Pauline Kael, and Anderson himself.

André Bazin, the dean of *Cahiers* critics, shows himself struggling to validate the line of younger critics of *Cahiers du Cinéma,* particularly that of Truffaut, who in the same issue devoted to Hitchcock (no. 39, October 1954) extravagantly claimed for Hitchcock's films a highly self-conscious metaphysical and moral structuring. Here in interviewing Hitchcock and examining his films, Bazin valuably redefines that line in more cinematic terms. So too do Robin Wood in his introduction to his book *Hitchcock's Films* (included here under the title "Why We Should Take Hitchcock Seriously") and Andrew Sarris in his short piece from *The American Cinema.* Finally Raymond Durgnat offers a more recent qualifying view of Hitchcock that takes account of the *Cahiers* approach without merely dismissing it as extravagance.

In part 3 "The Films," can be found detailed criticism of particular films. Again I have tried to offer a wide variety of approaches and films, but space was limited and in some cases material was not available. There is a dearth of criticism of the great British films of the 1930s; most of the critical enthusiasm has taken off from the *Cahiers* approach and centers around the American films. The English films need reexamination.

Except for excerpting his introduction, a key document in the Hitchcock controversy, I have avoided anthologizing from Robin Wood's *Hitchcock's Films.* The chapter on *Vertigo* is a masterpiece, but too long for inclusion here, and the book itself is easily available in paperback form. It is essential reading for Hitchcock enthusiasts, and in addition to the *Vertigo* chapter, I would strongly recommend the chapters on *Strangers on a Train* and *The Birds.* I have also avoided quoting the famous *Truffaut–Hitchcock* series of interviews; the volume is easily available and it seemed more valuable to print less easily available material here.

Some other important articles of Hitchcock criticism are missing. Most of them are from the English magazine *Movie* and represent a reworking of the *Cahiers* approach. I have high regard for V. F. Perkins study of Hitchcock's fluid camera work in *Rope* (*Movie* 7); though dealing with a lesser movie, it is the richest study of the implications of camera movement in Hitchcock that we have. Also extremely interesting—particularly in the light of objections made to the late films—are the detailed studies of *The Birds* and *Marnie* by Ian Cameron and Jeffery Richard, published under the title "The Universal Hitchcock" in *Movie* 12. Michael Walker's "The Old Age of Alfred Hitchcock" continues the defense brilliantly in *Movie* 18, this time with *Topaz*. I regret not having room to represent varying viewpoints on these late highly controversial films. Fortunately, all of the above articles will soon be available in an anthology of criticism on Hitchcock from *Movie* to be published by Praeger. Also forthcoming in a book titled *Godard by Godard,* translated by Tom Milne, is Godard's contribution to the *Cahiers* discussion on Hitchcock, a study of *The Wrong Man* entitled "The Cinema and Its Double."

The bibliography contains a list of most of the important books, interviews, and articles on Hitchcock, and my comments there should help to single out what is most interesting. Let me mention, however, three other articles for which there was no room, but which contain new approaches to Hitchcock that could be followed up: Molly Haskell's analysis of *Stage Fright* in *Film Comment* 6, no. 3 (Fall, 1970); Richard Corliss's qualifications about *Topaz* in *Film Quarterly* (Spring, 1970) and Charles Thomas Samuels's pursuit of the musical analogies with films in several of Hitchcock's films in *The American Scholar* (Spring 1970).

With the exception of these articles, the reader can find here most of the approaches to Hitchcock, the questions he raises, his own concerns, and those of his critics, and the intense excitement that his films generate even in discussion, once their immediate viewing excitement is over.

Chronology

1899	Born August 13 in London to William Hitchcock, a poultry dealer and fruit importer, and Emma (Whelan) Hitchcock.
1907	By age of eight had ridden every bus line in London; fascination with geography and maps; each day he buys the shipping bulletin and plots the location of British merchant fleet on wall map at home.
1912–20	Education: St. Ignatius College, a Jesuit school; University of London where he studies to be an electrical engineer (also art, navigation, economics, political science).
	First job: with a cable company making technical calculations on electrical systems to be installed by the firm.
	Abandons technology for art: job as assistant layout man in advertising office of London department store at fifteen shillings a week.
1920	Enters motion picture industry. Executive of Famous Players Lasky (now Paramount) comes to London planning to film *Sorrows of Satan*; H. arrives with portfolio only to learn that another film, *The Great Day* is planned. Overnight Hitchcock assembles another portfolio of titles and drawings and gets the job as title writer and title artist.
	Innovates in titles by adding symbolic drawings to them —a willingness to innovate and experiment with new techniques that he shows throughout his career. The title drawings also show his flair for compositional qualities that marks all his films.
1923	Joins Gainsborough Pictures in Islington, England, as scenario writer and worked with Graham Cutts on several films. Sometimes art director, assistant director, and production manager as well. Period of film apprenticeship.
1923	First film credit as art director of *Woman to Woman*.
1925	First film fully directed by Hitchcock, *The Pleasure Garden*.

18

1926   Marries Alma Reville, his assistant director on *The Pleasure Garden*. Throughout his career, she continues to provide continuity for his scripts, frequently writes the screenplays.

1926   First Hitchcock thriller, *The Lodger*, with Ivor Novello. Visual experimentation with the see-through ceiling.

1929   *Blackmail*, first made as a silent, then refilmed as a sound film. First British sound film. Innovations in use of sound, particularly in the hallucinatory "knife" repetitions.

1935–38   The Hitchcock cycle that establishes Hitchcock as England's great director: *The Man Who Knew Too Much* (1934), *The Thirty-Nine Steps* (1935), *The Secret Agent* (1936), *Sabotage* (1936), *Young and Innocent* (1937), and *The Lady Vanishes* (1938).

1938   Hitchcock wins best director award from New York Film Critics for *The Lady Vanishes*.

1938   First visit to the United States (though Hitchcock already knew all the train timetables and the geography).

1939   Return to United States and takes up permanent residence. Contract with Selznick.

1940   *Rebecca*, first American movie, wins Academy Award as best picture of the year. Hitchcock nominated for best director award.

1948   *Rope*, an experiment in continuous shooting, cutting only at the end of reels in the camera.

1950–60   Decade of great films: *Strangers on a Train, I Confess, Rear Window, To Catch a Thief, The Trouble with Harry, The Wrong Man, Vertigo, North by Northwest, Psycho*.

1951   *Strangers on a Train* marks for many Hitchcock fans a return to greatness associated with the English pictures. Beginning of association with Robert Burks as his cameraman for many films.

1954   Experiments with 3-D in filming *Dial M for Murder*

1954   *Cahiers du Cinéma* devotes an entire issue to Hitchcock as auteur. Articles by Truffaut, Bazin, Chabrol.

1955   Inaugurates *Alfred Hitchcock Presents* on TV.

1957   First full-length critical study by Eric Rohmer and Claude Chabrol.

1960–70   The late films over which there is more controversy: *The Birds, Marnie, Torn Curtain, Topaz*.

HITCHCOCK
ON
HITCHCOCK

I Wish I Didn't Have to Shoot the Picture: An Interview with Alfred Hitchcock

by BUDGE CRAWLEY, FLETCHER MARKLE, AND GERALD PRATLEY

Q: How have you managed to find the same challenge, stimulus, the inspiration? How do you continue to find something new and worthwhile to do as you go from picture to picture?

A: Well, I think that the main problem one has, in my particular field, is the avoidance of the cliché. You see, audiences now—with television, and having films for fifty years—are now highly educated in all forms of mayhem, crime: They're all experts—the public I mean. I was talking to a judge while I was making a film called *The Wrong Man* and he said that he wished they could have trials without juries, because juries were becoming—what is the word— something of a nuisance. They all want to know from the witness —if there's a police officer on the stand, they want to know, "What about the fingerprints, what about this, what about that?" They're all experts. So one has to recognize that you do have an audience today—with the increased facilities of communication, of television, films, paperbacks, and everything else—you have to be aware of this competition and meet it.

To give you an example of avoiding the cliché: I made a movie called *North by Northwest* and I had occasion to use a situation (which is a very old-fashioned one) of sending a man—in this case Cary Grant—to an appointed place: He's what they call "put on the spot." And there, probably, to be shot at. Now, the convention of this situation has been done many times: He is stood under the

street lamp at night in a pool of light, waiting, very sinister surroundings, the cobbles are all washed by the recent rain—you've
seen that in many pictures—then we cut to a window and a face
peers furtively out, then you cut to the bottom of the wall and a
black cat slithers along, then you wait for the limousine to arrive.
This is what we've been used to seeing. So, I decided, "I won't do it
that way"; I would do it in bright sunlight, not a nook or a cranny
or a corner of refuge for our victim. Now we have a situation where
the audience are wondering. A mad tension. And it's not going to
come out of a dark corner. So, not only do you give them suspense,
but you give them mystery as well. He's alone and then a man
arrives across the other side of the road, and he crosses to talk to
him and this man suddenly says, "Look, there's a crop duster over
there, dusting the field where there are no crops." Now, that's the
first thing that you give to the audience: this sinister, mysterious
comment. But, before it can be discussed, you put the man on the
bus and he drives off, so you and Cary Grant are now—because you
are identified with him—left alone. And then suddenly the airplane
comes down and shoots at him all over the place. . . . So there you
see an example of the very question you ask, "How do you keep up,
how do you change?" Only by rejecting the obvious and then, out
of that, you will find new ways to do the same thing.[1]

Q: You've been described, Mr. Hitchcock, not only as the master
of the horror film, but also as master of preplanned production
techniques. . . . How much improvisation is there in your films
and would you talk to us about your methods of filmmaking?

A: Well, in the first place I agree that you can improvise and
should improvise, but I think it should be done in an office, where
there are no electricians waiting and no actors waiting, and you can
improvise all you want—ahead of time. Sometimes, I compare it
with a composer who is trying to write a piece of music with a full
orchestra in front of him. Can you imagine him saying, "Flute, give
me that note again will you. Thank you, flute," and he writes it
down. . . . A painter has his canvas and he uses his charcoal sketch
and he goes to work on that canvas with a preconceived idea. I'm
sure he doesn't guess it as he goes along. So, I am not in approval of
the improvisation on the studio stage, while the actor is on the
phone about his next picture and all that kind of stuff.

Q: Mr. Hitchcock, how have you been able to resist, over fifty
years of direction, the temptation to look through the camera?

A: I don't look through the camera. Looking through the camera
has nothing to do with it. The ultimate end of what you're doing is

---

[1] [See also Robin Wood's analysis of this scene, this volume, pp. 81–82, and
the drawings and analysis, pp. 145–73.]

on a rectangular screen of varying proportions—wide ones, tall ones, all those kinds of screens—but, nevertheless, what are you doing? You're using the rectangle, like a painter, but the whole art of the motion picture is a succession of composed images, rapidly going through a machine, creating ideas. The average public do not, or are not, aware of "cutting" as we know it, and yet that is the pure orchestration of the motion-picture form. So, therefore, looking through a camera has absolutely nothing to do with it at all. It's the rectangle where the composition arrives. I would say, if I looked through a camera, having asked for a certain composition of a given set-up, it would be as though I distrusted the cameraman and he was a liar, and I'm testing him out.

Q: What about other directors?

A: I don't know anything about other directors; maybe I'm a snob in that direction. And I've never seen other directors at work; I never have. I've heard about them: They tell me they dress up for directing. I've always worn the same blue suit everywhere.

Q: What about seeing your rushes or your dailies? We hear that you pay little attention to them. Is this correct?

A: Yes, it is correct, because I go and check them up after about four or five days, but I don't rush the same evening to see, "Has it come out?" That would be like going to the local camera shop to see the snaps and make sure nobody has moved.

Q: Mr. Hitchcock, what about your editing methods? When do you start to edit your films, and are you able to edit them right through to the very end without anyone else interfering with it?

A: Well I—following what I have said—do shoot a precut picture. In other words, every piece of film is designed to perform a function. So therefore, literally, the only type of editing that I do is to tighten up. If a man's coming through the door, going into the room, then you just pull that together by just snippets. But actual creative work in the cutting, for me, is nonexistent, because it is designed ahead of time—precut, which it should be. You don't agree with me, huh?

Q: Oh yes.

A: Oh.

Q: Yes.

Q: Have you ever used a shot that, perhaps, might have been shot by accident on the set in a film of yours—that wasn't pre-planned?

A: Oh no, I don't think so. For example, in the film *Psycho,* I did a murder in the shower. I spent seven days on that—seventy-eight cuts for forty-five seconds of film. That meant you got pieces of film no bigger than two or three frames. And that was shot with the head of the leading lady; I had a nude girl—we shot a lot of her struggles —but more than that, what people don't realize in a situation like this, you had censor problems, so you had bare breasts to cover. So in order to measure this out, I had some parts of the scene shot in slow motion, so the girl moved like that to struggle and the arm covers the breasts there—which could never have been done had you shot it quickly because you couldn't measure it out.

Q: Have you been up to date, shall we say, in your new film *Torn Curtain?* Do you have any love scenes in this which . . . uh . . .

A: Oh yes, I have Julie Andrews and Newman in bed together discussing their wedding day. Although I must say—here's an ex-ample in this film, in this particular scene, of the avoidance of the cliché. I got so bored with seeing those English films with the nude couple in bed and that constant shot over the bare shoulder of the man, which is just covering the breasts of the girl—it's such a bore and so unimaginative that I took the trouble in the opening of *Torn Curtain* to show a ship in a Norwegian fjord and on board is an international convention of nuclear physicists, and I have turned the heat off on the ship. I have made the heat go wrong. The reason I did that was because I wanted all the people in the dining room to be wrapped in coats and freezing to death having their lunch. And then I go down below and show our couple in bed, covered with blankets, covered in topcoats, and you barely see them at all. For some inexplicable reason, my sense of propriety in this matter didn't seem to meet the approval of the Legion of Decency; they complained that there were premarital occupations going on, and I don't understand why they said that because I can't see a thing.

Q: Is the smallest period involved in production the shooting period?

A: Oh yes. I wish I didn't have to shoot the picture. When I've gone through the script and created the picture on paper, for me the creative job is done and the rest is just a bore.

. . . I think, to me, the great art of the motion picture is by means of imagery and montage to create an emotion in the au-dience and, therefore, the content is a means to an end. In other words, I would choose a story that would help toward that end rather than just photograph a story without any technique.

Q: Would you approve of a film which involved only technique and no story?

A: Oh yes, you have to have story because, you see, you need shape. You see, the nearest art form to the motion picture is, I think, the short story. It's the only form where you ask the audience to sit down and read it in one sitting.

In the film, you ask the audience to stay in one seat for two hours. Therefore, you need a shape of the story that has a rising curve of interest. You know, Bernard Shaw once tried to figure out how long an act of a play would be based on the endurance of the human bladder. And that is our fundamental problem when we devise a film. We do ask a person to sit there for two hours and therefore the shape and story-shape comes into it considerably because, as you get toward the end when they, your audience, might begin to be— shall we say—physically distracted, you must increase the interest on the screen to take their minds off this kind of thing.

Q: It's generally considered that films of mystery are best in black and white, yet you've photographed *Torn Curtain* in color. What have you been able to do with the color process and the photography which perhaps aids the mood you're trying to achieve?

A: Well, when they talk about black and white, you remember that black and white itself is unreal basically. After all, we see color everywhere. The camera will photograph whatever you give it. If you want to give it a black and white set—a woman in a black dress and a white blouse, there'll only be one thing in color; that'll be her face, the rest will all be black and white so that you can create the same thing in your own way. In *Torn Curtain,* we decided that after we leave Copenhagen, which is the last location in the picture before we go to East Germany, to go grey everywhere—grey and beige—so we have a mood, a depressed mood, a sinister mood, in the general tones of all the sets and they're all painted grey for that purpose. So you see black and white or color really don't have any relationship. The only reason I made a picture like *Psycho* in black and white is because of the amount of blood.

Q: Mr. Hitchcock, you're sixty-five and you directed *The Lady Vanishes* about '37. Do you find that it is just as easy now, in directing a picture like *Torn Curtain,* to keep up your enthusiasm as it was in the days when you were shooting *The Lady Vanishes,* and if not, why not?

A: Yes, you have to do that. After all, the most enjoyable part of making a picture is in that little office, with the writer, when we are discussing the story-lines and what we're going to put on the screen, searching for freshness and so forth, and also always that lovely moment when we say, "Wouldn't it be fun to kill him this way."

. . . The big difference is that I do not let the writer go off on his own and just write a script that I will interpret. I stay involved with him and get him involved in the direction of the picture. So he becomes more than a writer; he becomes part maker of the picture, because the picture is being made.

Q: If you were going to be murdered, how would you choose to have it done?

A: Well, there are many nice ways: Eating is a good one.

Q: Mr. Hitchcock, we have talked tonight a great deal about the technique of making motion pictures, we often hear a great deal said about the art of making motion pictures. Could you close by telling us just exactly where does the art come in. Is art technique or does the technique become art?

A: Well, I think that the art is in its basic form. The motion picture was the newest art form of the twentieth century and that is, its purest form, montage—pieces of film put together, shall we say, artfully, and creating ideas. But, you see, unfortunately, it's so little practiced today. We see so many films that are merely an extension of the theater: They are photographs of people saying lines and so forth. So I regret that enough films are not made using the pure art form.

Interviews with Alfred Hitchcock
by PETER BOGDANOVICH

1926   *The Lodger* (*A Story of the London Fog*)

Did you want the audience to believe without doubt that Novello was the murderer?

That was one of the commercial drawbacks one encountered. Of course, strictly speaking, he should have been the ripper and gone on his way. That's how Mrs. Belloc-Lowndes wrote the book. But Ivor Novello was the matinee idol of the period and could not be the murderer. The same thing was true of Cary Grant in *Suspicion* many years later. So, obviously, putting that kind of actor into this sort of film is a mistake because you just have to compromise.

In The Lodger *you were quite conscious of the German school of filmmaking, weren't you?*

Very much so. You have to remember that a year before, I was working on the Ufa lot—I worked there for many months, at the same time as Jannings was making *The Last Laugh* with Murnau. And I was able to absorb a lot of the methods and style.

How did you achieve the shot of Novello pacing back and forth above their heads?

I had a floor made of one-inch thick plate glass, about six feet square. This was the visual substitution for sound, you see. Just as much as the set I had built for when the lodger went out late at night—almost to the ceiling of the studio, showing four flights of stairs and a handrail. And all you see is a hand going down. That was, of course, from the point of view of the mother listening. Today we would substitute sound for that. Although I think that the handrail shot would be worthy of today in addition to sound.

From The Cinema of Alfred Hitchcock *by Peter Bogdanovich* (*New York: The Museum of Modern Art, 1963. Copyright © 1963 The Museum of Modern Art. Reprinted by permission of The Museum of Modern Art.*

1930   *Murder*

Hitchcock: "*Murder* was the first important who-done-it picture I made. It's the first time I ever used the voice over the face— without the lips moving—for stream-of-consciousness. Before O'Neill. And there was a scene where Marshall was shaving, and he had the radio on and I wanted to have the Prelude from *Tristan* playing. I had a thirty-piece orchestra in the studio, just for this little radio he's playing in his bathroom. You see, you couldn't add it later, it had to be done at the same time and balanced on the stage."

1935   *The Thirty-Nine Steps*

In all your chase films, why do you have the hero fleeing from both the police and the real criminals?

One of the reasons is a structural one. The audience must be in tremendous sympathy with the man on the run. But the basic reason is that the audience will wonder, "Why doesn't he go for the police?" Well, the police are after him, so he can't go to them, can he?

Isn't it his sense of guilt that makes him so fervent?

Well, yes, to some degree. In *Thirty-Nine Steps* maybe he feels guilt because the woman is so desperate and he doesn't protect her enough, he's careless.

Is The Thirty-Nine Steps *one of your favorite films?*

Yes. Pretty much. What I liked about *Thirty-Nine Steps* were the sudden switches and the jumping from one situation to another with such rapidity. Donat leaping out of the window of the police station with half of a handcuff on, and immediately walking into a Salvation Army Band, darting down an alleyway and into a room. "Thank God you've come, Mr. So-and-so," they say, and put him onto a platform. A girl comes along with two men, takes him in a car to the police station, but not really to the police station—they are two spies. You know, the rapidity of the switches, that's the great thing about it. If I did *The Thirty-Nine Steps* again, I would stick to that formula, but it really takes a lot of work. You have to use one idea after another, and with such rapidity.

1943   *Shadow of a Doubt*

Teresa Wright makes a lot out of the fact that she and her uncle are similar, and yet she is the most eager to suspect him of the worst.

Only because her attention is drawn to him more than anybody else. You look at your adoring uncle long enough, and you find something.

Isn't Cotten rather sympathetic in the film?

There is sympathy for any murderer, or let's call it compassion. You hear of murderers who feel they've been sent to destroy. Maybe those women deserved what they got, but it wasn't *his* job to do it. There is a moral judgment—he is destroyed at the end, isn't he? The girl unwittingly kills her own uncle. She is the instrument by which he falls in front of the train. It comes under the heading that all villains are not black and all heroes are not white. There are grays everywhere.

Does Cotten really love Wright in the film?

I don't think so. Not as much as she loves him. And yet she destroys him. She has to. Wasn't it Oscar Wilde who said, "You destroy the thing you love?" *Shadow of a Doubt* was a most satisfying picture for me—one of my favorite films—because for once there was time to get characters into it. It was the blending of character and thriller at the same time. That's very hard to do.

1951   *Strangers on a Train*

Hitchcock: "Granger was miscast. Warners insisted I take him. It should have been a much stronger man. The stronger the man, the more frustrated he would have been in the situation."

Isn't the irony of the picture that Walker actually does free Granger from his impossible wife?

Sure. Granger didn't pay back, did he? He didn't kill Walker's father. He ratted on Walker.

How did you achieve that stunning carousel sequence?

This was a most complicated sequence. For rear-projection shooting there is a screen and behind it is an enormous projector throwing an image on the screen. On the studio floor is a narrow white line right in line with the projector lens and the lens of the camera must be right on that white line. That camera is not photographing the screen and what's on it, it is photographing light in certain colors, therefore the camera lens must be level and in line with the projector lens. Many of the shots on the merry-go-round were low camera set-ups. Therefore you can imagine the problem. The projector had to be put up on a high platform, pointing down, and the screen had to be exactly at right angles to the level-line from the lens. All the shots took nearly half a day to line up, for each set-up. We had to change the projector every time the angle changed. When the carousel broke, that was a miniature blown up on a big screen and we put live people in front of the screen. But I did the most dangerous thing I've ever done in that picture and I'll never, never do it again. When the little man crawled underneath the moving carousel—that was actual. If he had raised his head an inch, two inches—finish. My hands sweat now when I think of it— what a dreadful chance I took. I knew what I was doing then, you know, but I thought, "Oh, well, maybe he won't raise his head too high."

Doesn't Granger chase after Walker mainly to expiate his own feelings of guilt about the murder of his wife?

Sure he does. He felt like killing her himself.

Direction (1937)
by ALFRED HITCHCOCK

Many people think a film director does all his work in the studio, drilling the actors, making them do what he wants. That is not at all true of my own methods, and I can write only of my own methods. I like to have a film complete in my mind before I go on the floor. Sometimes the first idea one has of a film is of a vague pattern, a sort of haze with a certain shape. There is possibly a colorful opening developing into something more intimate; then, perhaps in the middle, a progression to a chase or some other adventure; and sometimes at the end the big shape of a climax, or maybe some twist or surprise. You see this hazy pattern, and then you have to find a narrative idea to suit it. Or a story may give you an idea first and you have to develop it into a pattern.

Imagine an example of a standard plot—let us say a conflict between love and duty. This idea was the origin of my first talkie, *Blackmail*. The hazy pattern one saw beforehand was duty—love—love versus duty—and finally either duty or love, one or the other. The whole middle section was built up on the theme of love versus duty, after duty and love had been introduced separately in turn. So I had first to put on the screen an episode expressing duty.

I showed the arrest of a criminal by Scotland Yard detectives, and tried to make it as concrete and detailed as I could. You even saw the detectives take the man to the lavatory to wash his hands—nothing exciting, just the routine of duty. Then the young detective says he's going out that evening with his girl, and the sequence ends, pointing on from duty to love. Then you start showing the relationship between the detective and his girl: They are middle-class people. The love theme doesn't run smoothly; there is a quarrel and the girl goes off by herself, just because the young man has kept her waiting a few minutes. So your story starts; the girl falls in with

From Charles Davy, ed., Footnotes to the Film *(London: Lovat Dickson and Thompson, Ltd., 1937). Reprinted by permission of Peter Davies Limited.*

the villain—he tries to seduce her and she kills him. Now you've got your problem prepared. Next morning, as soon as the detective is put onto the murder case, you have your conflict—love versus duty. The audience know that he will be trying to track down his own girl, who has done the murder, so you sustain their interest: They wonder what will happen next.

The blackmailer was really a subsidiary theme. I wanted him to go through and expose the girl. That was my idea of how the story ought to end. I wanted the pursuit to be after the girl, not after the blackmailer. That would have brought the conflict on to a climax, with the young detective, ahead of the others, trying to push the girl out through a window to get her away, and the girl turning round and saying: "You can't do that—I must give myself up." Then the rest of the police arrive, misinterpret what he is doing, and say, "Good man, you've got her," not knowing the relationship between them. Now the reason for the opening comes to light. You repeat every shot used first to illustrate the duty theme, only now it is the girl who is the criminal. The young man is there ostensibly as a detective, but of course the audience know he is in love with the girl. The girl is locked up in her cell and the two detectives walk away, and the older one says, "Going out with your girl to-night?" The younger one shakes his head. "No. Not tonight."

That was the ending I wanted for *Blackmail*, but I had to change it for commercial reasons. The girl couldn't be left to face her fate. And that shows you how the films suffer from their own power of appealing to millions. They could often be subtler than they are, but their own popularity won't let them.

But to get back to the early work on a film. With the help of my wife, who does the technical continuity, I plan out a script very carefully, hoping to follow it exactly, all the way through, when shooting starts. In fact, this working on the script is the real making of the film, for me. When I've done it, the film is finished already in my mind. Usually, too, I don't find it necessary to do more than supervise the editing myself. I know it is said sometimes that a director ought to edit his own pictures if he wants to control their final form, for it is in the editing, according to this view, that a film is really brought into being. But if the scenario is planned out in detail, and followed closely during production, editing should be easy. All that has to be done is to cut away irrelevancies and see that the finished film is an accurate rendering of the scenario.

Settings, of course, come into the preliminary plan, and usually I have fairly clear ideas about them; I was an art student before I took up with films. Sometimes I even think of backgrounds first. *The Man Who Knew Too Much* started like that; I looked in my mind's eye at snowy Alps and dingy London alleys, and threw my characters into the middle of the contrast. Studio settings, however, are often a problem; one difficulty is that extreme effects—extremes of luxury or extremes of squalor—are much the easiest to register on the screen. If you try to reproduce the average sitting room in Golders Green or Streatham it is apt to come out looking like nothing in particular, just nondescript. It is true that I have tried lately to get interiors with a real lower-middle-class atmosphere—for instance, the Verlocs' living room in *Sabotage*—but there's always a certain risk in giving your audience humdrum truth.

However, in time the script and the sets are finished somehow and we are ready to start shooting. One great problem that occurs at once, and keeps on occurring, is to get the players to adapt themselves to film technique. Many of them, of course, come from the stage; they are not cinema-minded at all. So, quite naturally, they like to play long scenes straight ahead. I am willing to work with the long uninterrupted shot: You can't avoid it altogether, and you can get some variety by having two cameras running, one close up and one farther off, and cutting from one to the other when the film is edited. But if I have to shoot a long scene continuously I always feel I am losing grip on it, from a cinematic point of view. The camera, I feel, is simply standing there, *hoping* to catch something with a visual point to it. What I like to do always is to photograph just the little bits of a scene that I really need for building up a visual sequence. I want to put my film together on the screen, not simply to photograph something that has been put together already in the form of a long piece of stage acting. This is what gives an effect of life to a picture—the feeling that when you see it on the screen you are watching something that has been conceived and brought to birth directly in visual terms. The screen ought to speak its own language, freshly coined, and it can't do that unless it treats an acted scene as a piece of raw material which must be broken up, taken to bits, before it can be woven into an expressive visual pattern.

You can see an example of what I mean in *Sabotage*. Just before Verloc is killed there is a scene made up entirely of short pieces of film, separately photographed. This scene has to show how Verloc comes to be killed—how the thought of killing him arises in Sylvia Sidney's mind and connects itself with the carving knife she uses when they sit down to dinner. But the sympathy of the audience has to be kept with Sylvia Sidney; it must be clear that Verloc's death, finally, is an accident. So, as she serves at the table, you see her unconsciously serving vegetables with the carving knife, as though her hand were keeping hold of the knife of its own accord. The camera cuts from her hand to her eyes and back to her hand; then back to her eyes as she suddenly becomes aware of

the knife making its error. Then to a normal shot—the man un-
concernedly eating; then back to the hand holding the knife. In
an older style of acting Sylvia would have had to show the audience what was passing in her mind by exaggerated facial expression. But people today in real life often don't show their feelings
in their faces, so the film treatment showed the audience her mind
through her hand, through its unconscious grasp on the knife. Now
the camera moves again to Verloc—back to the knife—back again
to his face. You see him seeing the knife, realizing its implication.
The tension between the two is built up with the knife as its focus.

Now when the camera has immersed the audience so closely in
a scene such as this, it can't instantly become objective again. It
must broaden the movement of the scene without loosening the
tension. Verloc gets up and walks round the table, coming so close
to the camera that you feel, if you are sitting in the audience,
almost as though you must move back to make room for him.
Then the camera moves to Sylvia Sidney again, then returns to
the subject—the knife.

So you gradually build up the psychological situation, piece by
piece, using the camera to emphasize first one detail, then another.
The point is to draw the audience right inside the situation instead of leaving them to watch it from outside, from a distance.
And you can do this only by breaking the action up into details
and cutting from one to the other, so that each detail is forced
in turn on the attention of the audience and reveals its psychological meaning. If you played the whole scene straight through,
and simply made a photographic record of it with the camera
always in one position, you would lose your power over the audience. They would watch the scene without becoming really involved in it, and you would have no means of concentrating their
attention on those particular visual details which make them feel
what the characters are feeling.

This way of building up a picture means that film work hasn't
much need for the virtuoso actor who gets his effects and climaxes
himself, who plays directly on to the audience with the force
of his talent and personality. The screen actor has got to be
much more plastic; he has to submit himself to be used by the
director and the camera. Mostly he is wanted to behave quietly
and naturally (which, of course, isn't at all easy), leaving the camera
to add most of the accents and emphases. I would almost say that
the best screen actor is the man who can do nothing extremely well.

One way of using the camera to give emphasis is the reaction
shot. By the reaction shot I mean any close-up which illustrates an
event by showing instantly the reaction to it of a person or a
group. The door opens for someone to come in, and before showing

who it is you cut to the expressions of the persons already in the room. Or, while one person is talking, you keep your camera on someone else who is listening. This overrunning of one person's image with another person's voice is a method peculiar to the talkies; it is one of the devices which help the talkies to tell a story faster than a silent film could tell it, and faster than it could be told on the stage.

Or, again, you can use the camera to give emphasis whenever the attention of the audience has to be focused for a moment on a certain player. There is no need for him to raise his voice or move to the center of the stage or do anything dramatic. A close-up will do it all for him—will give him, so to speak, the stage all to himself.

I must say that in recent years I have come to make much less use of obvious camera devices. I have become more commercial-minded, afraid that anything at all subtle may be missed. I have learnt from experience how easily small touches are overlooked. The other day a journalist came to interview me and we spoke about film technique. "I always remember," he said, "a little bit in one of your silent films, *The Ring*. The young boxer comes home after winning his fight. He is flushed with success—wants to celebrate. He pours out champagne all round. Then he finds that his wife is out, and he knows at once that she is out with another man. At this moment the camera cuts to a glass of champagne; you see a fizz of bubbles rise off it and there it stands untasted, going flat. That one shot gives you the whole feeling of the scene." Yes, I said, that sort of imagery may be quite good: I don't despise it and still use it now and then. But is it always noticed? There was another bit in *The Ring* which I believe hardly any one noticed.

The scene was outside a boxing booth at a fair, with a barker talking to the crowd. Inside the booth a professional is taking on all comers. He has always won in the first round. A man comes running out of the booth and speaks to the barker: Something un-expected has happened. Then a cut straight to the ringside: You see an old figure 1 being taken down and replaced by a brand new figure 2. I meant this single detail to show that the boxer, now, is up against someone he can't put out in the first round. But it went by too quickly. Perhaps I might have shown the new figure 2 being taken out of a paper wrapping—something else was needed to make the audience see in a moment that the figure for the second round had never been used before.

The film always has to deal in exaggerations. Its methods reflect the simple contrasts of black-and-white photography. One advan-tage of color is that it would give you more intermediate shades. I should never want to fill the screen with color: It ought to be used economically—to put new words into the screen's visual language when there's a need for them. You could start a color film with a board-room scene: somber paneling and furniture, the directors all in dark clothes and white collars. Then the chairman's wife comes in, wearing a red hat. She takes the attention of the audience at once, just because of that one note of color. Or suppose a gangster story: The leader of the gang is sitting in a cafe with a man he suspects. He has told his gunman to watch the table. "If I order a glass of port, bump him off. If I order green chartreuse, let him go."

This journalist asked me also about distorted sound—a device I tried in *Blackmail* when the word "knife" hammers on the consciousness of the girl at breakfast on the morning after the murder. Again, I think this kind of effect may be justified. There have always been occasions when we have needed to show a phantasmagoria of the mind in terms of visual imagery. So we may want to show someone's mental state by letting him listen to some sound —let us say church bells—and making them clang with distorted insistence in his head. But on the whole nowadays I try to tell a story in the simplest possible way, so that I can feel sure it will hold the attention of any audience and won't puzzle them. I know there are critics who ask why lately I have made only thrillers. Am I satisfied, they say, with putting on the screen the equivalent merely of popular novelettes? Part of the answer is that I am out to get the best stories I can which will suit the film medium, and I have usually found it necessary to take a hand in writing them myself.

There is a shortage of good writing for the screen, and is that surprising? A playwright may take a year or more writing a play, but in a year the film industry has to make hundreds of films. More and more pictures, one after the other incessantly, with a certain standard to keep up—it throws a great strain on the creative faculties of everyone who has to supply the industry with ideas. Of course there must be cooperation, division of labor, all the time. The old saying, "No one man ever made a picture," is entirely true. And the only answer found so far to the writing problem has been to employ a number of writers to work together on the same picture. Metro-Goldwyn, we are told, employ altogether a staff of eighty or ninety writers, so they can draw at any time on a whole group of writers to see a story through. I don't say there aren't drawbacks in this collective method, but it often makes things easier when time is at stake, as it always is in film production. In this country we can't usually afford to employ large writing staffs, so I have had to join in and become a writer myself. I choose crime stories because that is the kind of story I can write,

or help to write, myself—the kind of story I can turn most easily
into a successful film. It is the same with Charles Bennett, who has
so often worked with me; he is essentially a writer of melodrama.
I am ready to use other stories, but I can't find writers who will
give them to me in a suitable form.

Sometimes I have been asked what films I should make if I were
free to do exactly as I liked without having to think about the box
office. There are several examples I can give very easily. For one
thing, I should like to make travel films with a personal element
in them: That would be quite a new field. Or I should like to do
a verbatim of a celebrated trial—of course there would have to be
some editing, some cutting down. The Thompson-Bywaters case,
for instance. You can see the figures at Madame Tussaud's and
the newspapers gave long reports of the trial. The cinema could
reconstruct the whole story. Or there is the fire at sea possibility
—that has never been tackled seriously on the screen. It might be
too terrifying for some audiences but it would make a great subject,
worth doing.

British producers are often urged to make more films about characteristic phases of English life. Why, they are asked, do we see so
little of the English farmer or the English seaman? Or is there not
plenty of good material in the great British industries—in mining
or shipbuilding or steel? One difficulty here is that English audiences seem to take more interest in American life—I suppose because it has a novelty value. They are rather easily bored by everyday scenes in their own country. But I certainly should like to
make a film of the Derby, only it might not be quite in the popular
class. It would be hard to invent a Derby story that wasn't hack-
neyed, conventional. I would rather do it more as a documentary
—a sort of pageant, an animated modern version of Frith's *Derby
Day*. I would show everything that goes on all round the course,
but without a story.

Perhaps the average audience isn't ready for that, yet. Popular
taste, all the same, does move; today you can put over scenes that
would have been ruled out a few years ago. Particularly towards
comedy, nowadays, there is a different attitude. You can get comedy
out of your stars, and you used not to be allowed to do anything
which might knock the glamor off them.

In 1926 I made a film called *Downhill*, from a play by Ivor
Novello, who acted in the film himself, with Ian Hunter and Isabel
Jeans. There was a sequence showing a quarrel between Hunter
and Novello. It started as an ordinary fight; then they began throwing things at one another. They tried to pick up heavy pedestals to
throw and the pedestals bowled them over. In other words I made
it comic. I even put Hunter into a morning coat and striped

trousers because I felt that a man never looks so ridiculous as when he is well dressed and fighting. This whole scene was cut out; they said I was guying Ivor Novello. It was ten years before its time.

I say ten years, because you may remember that in 1936 MGM showed a comedy called *Libeled Lady*. There is a fishing sequence in it: William Powell stumbles about in the river, falls flat and gets soaked, and catches a big fish by accident. Here you have a star, not a slapstick comedian, made to do something pretty near slapstick. In *The Thirty-Nine Steps*, too, a little earlier, I was allowed to drag Madeleine Carroll over the moors handcuffed to the hero; I made her get wet and untidy and look ridiculous for the purpose of the story. I couldn't have done that ten years ago.

I foresee the decline of the individual comedian. Of course there may always be specially gifted comedians who will have films written round them, but I think public taste is turning to like comedy and drama more mixed up; and this is another move away from the conventions of the stage. In a play your divisions are much more rigid; you have a scene—then curtain, and after an interval another scene starts. In a film you keep your whole action flowing; you can have comedy and drama running together and weave them in and out. Audiences are much readier now than they used to be for sudden changes of mood; and this means more freedom for a director. The art of directing for the commercial market is to know just how far you can go. In many ways I am freer now to do what I want to do than I was a few years ago. I hope in time to have more freedom still—if audiences will give it to me.

REAR WINDOW
by ALFRED HITCHCOCK

I chose this picture because of all the films I have made, this to me is the most cinematic. I'm a purist so far as the cinema is concerned. You see many films that are what I call photographs of people talking. This film has as its basic structure the purely visual. The story is told only in visual terms. Only a novelist could do the same thing. It's composed largely of Mr. Stewart as a character in one position in one room looking out onto his courtyard. So what he sees is a mental process blown up in his mind from the purely visual. It represents for me the purest form of cinema which is called montage: that is, pieces of film put together to make up an idea.

When the film was originally invented, when cutting was invented, it was the juxtaposition of pieces of film that went through a machine that displayed ideas on the screen. Unfortunately today a lot of that is lost: It's not being used sufficiently, or sometimes not at all. I think it was Pudovkin, the famous Russian director many years ago, who took a close-up and he put various objects in front of a woman's face and it was the combination of her face—she never changed her expression—and what she looked at (whether it was food or a child or what have you) that seemed to give an expression to her face. I made up the whole of the film production section of the *Encyclopedia Britannica* and I took the idea of this film as a prime example of the power of montage. For example, if Mr. Stewart is looking out into this courtyard and—let's say—he sees a woman with a child in her arms. Well, the first cut is Mr. Stewart, then what he sees, and then his reaction. We'll see him smile. Now if you took away the center piece of film and substituted —we'll say—a shot of the girl Miss Torso in a bikini, instead of being a benevolent gentleman he's now a dirty old man. And you've only changed one piece of film, you haven't changed his look or his

From Take One 2, no. 2 (*November–December 1968*): 18–20. Copyright © 1968 by Unicorn Publishing Corporation. Reprinted by permission of the publisher.

reaction. This is one of the reasons why I chose this film. You see, many people think that a little dialogue scene in a movie is motion pictures. It's not. It's only part of it. Galloping horses in Westerns are only photographs of action, photographs of content. But it's the piecing together of the montage which makes what I call a pure film.

In *Vertigo* and other of my pictures a lot of the visual "pure cinema" techniques are used, but the subject matter is the thing that lends itself to certain treatments. I use the cinematic technique as often as I can, but sometimes there isn't the opportunity. Certainly I think that this film of all of them presented the greatest opportunity.

I have made films based on stage plays (back in the very early days of talking pictures) where I found that when filming a stage play, it's best not to, what they call in our business "open it up," because a stage play is designed for a limited area of presentation, that is, the proscenium arch. Some years ago I tried to get around this problem when I made a film called *Rope*. It was a stage play and it played continuously in its own time. And I tried to give it a flowing camera movement and I didn't put any cuts in at all. I tried to do it as if I were giving the audience all opera glasses to follow the action on the stage, but basically it was on stage. I think people make a dreadful error when they "open up" stage plays. What do they do when they say "open it up"? Well they open it up with a shot of Fifth Avenue and a Yellow Cab pulls up, the characters come out of the cab, they cross the sidewalk, they go into the building, they press the button for the elevator, they go up, they get out of the elevator, they go around along a corridor, they press another button, and when the door opens, where are we? Back on the stage.

For *Rear Window* each cut was written ahead of time. It's like scoring music—I prefer to make a film on paper. People ask me, "Don't you ever improvise on the set?" and I say "No, I prefer to improvise in the office while we're writing. That's where the ideas come from." So I prefer to design this kind of film well ahead of time, with each cut in its proper place. It's like composing. A lot of films are made where they have a first draft script and make it up as they go along. To me that's like a composer trying to compose music with an orchestra in front of him. He has a blank sheet and he says, "Flute, give me a note will you." So I work strictly on paper.

There's no score in *Rear Window*. I was a little disappointed at the lack of a structure in the title song. I had a motion-picture songwriter when I should have chosen a popular songwriter. I was rather hoping to use the genesis, just the idea of a song which would then gradually grow and grow until it was used by a full orchestra. But I don't think that came out as strongly as I would have liked it to have done.

Rear Window has a happy ending, but I don't think you have to drag in a happy ending. I think that an audience will accept any ending as long as it's reasonable. Years ago I made a film of Sean O'Casey's *Juno and the Paycock*. It has a tragic ending, a very grim ending, but there was no other way around it. *Vertigo* ended with a girl falling from a tower in the same manner that she had helped a murderer with previously. The ending depends really on the nature of the content of what has gone before. Sometimes if you've created a lot of suspense in an audience it's very essential that you relieve that tension at the very end.

The rhythm of the cutting in *Rear Window* speeds up as the film goes on. This is because of the nature of the structure of the film. At the beginning, life is going on quite normally. The tempo is leisurely. There's a bit of a conflict between the man and the girl. And then gradually the first suspicion grows and it increases. And naturally as you reach the last third of your picture the events have to pile on top of each other. If you didn't, and if you slowed the tempo down, it would show up considerably. In the film *Psycho,* you start off with just a sack of money and a girl who is suddenly murdered in a shower. The shower scene was made very violent because of what was to follow. The pattern there was that events again increased, but I'd decreased the violence because I'd transferred the violence from the screen to the mind of the audience. So I didn't have to be violent later on because I'd built up the apprehension—having given them a sample, shall we say, and so it was a matter of going on and on increasing your tempo of events but keeping the violence down and letting the audience carry that for you, you see.

When you come down to the question of color, again it's the same as the orchestration with cutting. If you noticed in *Rear Window,* Miss Lonely Hearts always dressed in emerald green. To make sure that that came off, there was no other green in the picture, because we had to follow her very closely when she went across the street into the cafe. So I reserved that color for her. In *Dial M for Murder* I had the woman dressed in red to begin with and as the tragedy overtook her she went to brick, then to grey, then to black.

Since my scripts are worked out beforehand, there is no opportunity for creative work on the part of the film editor. I don't mind the film editor being in . . . well, in fact, even with the writer I let him be part of the direction of the picture. Working closely with the writer I can tell him how we're going to shoot it, what size image, and so forth. So I'm willing to share the creative end of it with the writer and the same would apply with the editor. But, you see, where the work of the average editor comes in is when he's given a lot of film to sort out. This is when directors use many angles of the same scene. But I never do that. As a matter of fact when this film, *Rear Window,* was finished somebody went into the cutting room and said, "Where are the out-takes? Where is the unused film?" And there was a small roll of a hundred feet. That was all that was left over.

If you want to be really mean towards the character in this film you could call him a Peeping Tom. I don't think it's necessarily a statement of morality because it's a statement of fact. You don't hide from it, there's no point in my leaving it out. When Grace Kelly says that they're a couple of fiendish ghouls because they're disappointed that a murder hasn't been committed she's speaking the truth. They were a couple of ghouls.

The MacGuffin in this story is really the wedding ring, which is the clue. The MacGuffin is really a nickname for what happens in spy stories. Or it's the papers that are stolen. It's something that the characters in the film care a lot about, but the audience doesn't worry about it too much. It's the plan for the fort or what have you. In Rudyard Kipling it could always be the Khyber Pass and the forts around it. Years ago I made a film called *Thirty-Nine Steps* and someone said, well what were the spies after, and it turned out to be a lot of gibberish which nobody . . . it was an airplane engine or bomb-bay door or something. As a matter of fact I refuse to use the kind of thing which most people think is very important. In the picture *North by Northwest,* Cary Grant speaking of the heavy or the spy says to the CIA, "Well, what is the fellow after?" and they answer, "Well, let's say he's an importer-exporter." And Grant says, "But what of?" and they answer "Government secrets." And that's all that was needed. The word MacGuffin comes from a story about two men in an English train, and one says to the other "What's that package on the baggage rack over your head?" "Oh," he says, "that's a MacGuffin." The first one says, "Well, what's a MacGuffin?" "It's an apparatus for trapping lions in the Scottish highlands." So the other says, "But there are no lions in the Scottish highlands." And he answers, "Then that's no

MacGuffin." To show you how people do make a big mistake about this kind of thing, I once designed a picture with Ben Hecht. It was called *Notorious*. And it dealt with the sending of a woman, Ingrid Bergman, and an agent down to Rio to see what some Nazis were up to. They were up to something. So the producer said, "Well, what are you going to have the Nazis doing down in Rio?" And I said, "Well, I thought that we were going to have them searching for samples of uranium 235." And he said, "What's that?" And I said—this is 1944—"Well, that's the stuff they're going to make the atom bomb out of." And he said, "What atom bomb? I've never heard of it." And I said, "No, it isn't out yet." As a result of me making this mistake, the producer didn't believe a word I said and finally sold the project to another studio, for only 50 percent of the property. He could have made 100 percent had he not made that cardinal error. Then I did meet some producers years after who said, "You know, we were offered a story of yours, *Notorious,* and we thought that was the Goddamndest thing on which to base a picture. How did you know years before it happened?" I said, "Well, there were all kinds of rumors. The Germans were dealing with heavy water in Norway." And so those producers lost all kinds of money for the wrong kind of thinking. But they still think that way. They still think that if the film's a spy film that it's all about . . . well, the MacGuffin.

On the question of violence, you see, you've got to go right back to the three-month-old baby. He's held in his mother's arms and the mother says to him, "Boo!" And the mother is being violent. And the child gets the hiccups and then the child smiles and the mother is very pleased with what she's done. It starts as early as that. In other words she scares the hell out of the baby. And that's how fear is born. And later the child grows up and goes on a swing and becomes violent to itself. It goes higher, and higher, and higher, and then it goes over the top. And next it tries a new kind of violence by going to the midway and going on the roller coaster, and then it goes shooting at rifle ranges, and knocking down objects. And the child is forced to read Hans Anderson or Grimm—you'll notice the word "grim." They take the child to see Hansel and Gretel and how they push an old woman into the oven. So there's nothing new in it. We've always had violence—it's communication. We've always had violence. We didn't have television, we didn't have radio years ago, but the violence was always there. Little boys point at each other and say "Bang, you're dead." And the other little boy rolls over. They don't believe it. People are fearful that children who are brought up to look at movies and television are violent. It isn't true. A little boy once came up to me and said "Oh, Mr. Hitchcock, in that scene in *Psycho,* what did you use for blood, chickens' blood?" And I said, "No. Chocolate sauce." But he said, "What did you use?" So I'm not sure that all the hullabaloo about violence is really correct. So far as the average individual child seeing movies, seeing Westerns with horses rolling over and bodies falling. . . . It reminds me also, going back to *Psycho,* I had a call from the Los Angeles *Times.* One of the reporters said, "A man has just been arrested for the murder of three women. And he confessed to murdering the third woman after seeing *Psycho.*" What did I have to say about it? I said, "What film did he see before he murdered the second woman? And am I to assume that the first woman was murdered after he had just finished drinking a glass of milk?"

The delineation of suspense covers a very, very wide field. Basically it is providing the audience with information that the characters do not have. The most simple example, the elementary example, is if four men are seated around a table and they're having a discussion about baseball, anything you like. Suddenly a bomb goes off and blows everyone to smithereens. Now, the audience get from that fifteen seconds of shock. But up to that time you've spent five minutes on a conversation about baseball. And the audience are without any knowledge that that bomb is under the table. Now let's take it the other way around. We show the bomb under the table, and let the audience know it's going to go off in five minutes. Now you go on with your conversation. Now the conversation becomes very potent, with the audience saying, "Stop talking about baseball, there's a bomb under there." Just as in *Rear Window* people were anxious about Grace Kelly being in the room and the man coming along the corridor. You're giving them information that neither of the characters have. So now you know there's a bomb under there and at the end of five minutes it's about to go off. You've driven the audience to the point of anxiety. Now a foot must touch the bomb and someone must look under, discover there's a bomb, pick it up, and throw it out the window. But it mustn't go off under the table. Because if you create suspense in the audience, it needs to be relieved of that suspense. Now, I made a film years ago from a Joseph Conrad story called *Secret Agent.* And I had a scene where a small boy carries a package across London. And he didn't know, but the audience knew, that a bomb was inside. And it had to be left at a certain place at a certain time. Well. I showed every form of holdup. He was even held up by the Lord Mayor's procession. Then he got onto a slow-moving bus, stop signs, go signs, policemen. And it drove the audience crazy. And I'd told them one o'clock and then I let it go on to one minute past one, two minutes past one. And at four minutes past one the bomb went off and blew up the bus, the boy into little pieces. I'd committed a cardinal sin. I had let that bomb go off. People were furious, angry.

I remember at the press show the leading London press woman critic came up and nearly hit me. "How dare you do a thing like that!" I hadn't relieved the suspense.[1]

Yes, I saw Truffaut's *The Bride Wore Black*. I thought it was quite well done. Well, people said that it was a tribute to me. The only thing that bothered me was I didn't know how the woman got to know that there were five men up in that room. But maybe he was getting mixed up with the MacGuffin.

---

[1] [Hitchcock's film of the Conrad novel is called *Sabotage* (1936). (His movie *The Secret Agent* (1936) is from Somerset Maugham's novel *Ashenden*.) The story of "not relieving the suspense" here is one of his favorites. Most people would feel that the suspense is relieved, but they are angry at Hitchcock for killing the boy and the people in the bus. Perhaps this lingering objection to the incident qualifies as not relieving suspense to Hitchcock. Curiously, his memory of the scene is somewhat incorrect in details, but accurate in regard to the effect. The bomb is not delayed as long as he says; it goes off on time at the last possible elongated second. But by cutting from clock to people, dogs, boy, and package, Hitchcock has made it seem as though the bomb is not going to go off.]

HITCHCOCK CONTROVERSY

Alfred Hitchcock
by LINDSAY ANDERSON

As, geographically, Britain is poised between continents, not quite Europe, and very far from America, so from certain points of view the British cinema seems to hover between the opposite poles of France and Hollywood. Our directors and producers never—or rarely—have the courage to tackle, in an adult manner, the completely adult subject; yet they lack also the flair for popular showmanship that is characteristic of the American cinema. It is significant that the most widely celebrated of all British directors should be remarkable for just this quality. So much so indeed that, when his powers were at their prime, he emigrated to Hollywood; and today, when he returns to work again in a British studio, he carries with him the pervasive aura of Hollywood success, and stays at the Savoy Hotel.

Alfred Hitchcock's long career has been intimately bound up with the history of the cinema. He began in the early twenties, title writing, then joined Michael Balcon's first production company, first as assistant and art director, then directing on his own. Between 1925 and 1929 he made nine pictures, and established himself as the foremost British director of the day. His *Blackmail* was the first British sound film. During the thirties he went on to perfect his grasp of technique, win a Hollywood contract and the opportunity to exploit the finest technical resources in the world. Essentially he is a man of the cinema—one who has approached the film as an art through the film as an industry.

His first two films are remarkable for their evidence of an immediate ease, an instinctive facility in the medium. *The Pleasure Garden* (1925) is a novelettish story—a good-hearted chorus girl befriends a vixenish young dancer, and ends up eight reels later menaced by her drunken husband, who believes himself incited to murder her by the ghost of his native mistress (whom he has drowned

From Sequence *9 (Autumn 1949). Copyright 1949 by Lindsay Anderson. Reprinted by permission of the author.*

in the lagoon). The most enjoyable passages are at the start: the first shots of chorus girls hurrying down a circular iron stair, then out on to the stage, gyrating enthusiastically in the abandoned fashion of the period. *The Mountain Eagle,* another romantic melodrama, was set equally far from home, among the hillbillies of Kentucky; one is not surprised to find *The Bioscope* commenting that "in spite of skillful and at times brillian direction, the story has an air of unreality."

Both these films were produced by Balcon in Munich; in 1926 Hitchcock returned to Islington to make his first picture in Britain, and the first opportunity to work on the sort of subject most congenial to him—the story of uncertainty, suspense, and horror amid humdrum surroundings. *The Lodger* was again a melodrama, but biased this time towards violence rather than romance. One winter evening, in a London terrorized by a homicidal maniac known as The Avenger, a handsome stranger arrives at a Bloomsbury lodging-house. He behaves strangely, creeping from the house at night, removing from his wall the portrait of a beautiful fair-haired girl (The Avenger attacks only blondes), and gradually the suspicion is built up that he is the Avenger himself. *The Lodger* is by no means a perfect thriller; it creates its suspense too often illegitimately: the innocent young man behaves like a stage villain, arriving out of the night heavily muffled and mysteriously silent. Playing chess before the fire with his landlady's attractive (blonde) daughter, he remarks with sinister emphasis, "Be careful, I'll get you yet," and picks up the poker—only to poke the fire vigorously on the entrance of a third person into the room.

This improbable development of the plot is partially disguised by the conscientious realism of its locales and characters: the authentic middle-class decors and homely atmosphere of the Buntings' house in Bloomsbury, the mannequins' dressing room at the couturier's where Daisy works, the flirtatious progress of Daisy's affair with Joe, the detective in charge of the case. Most remarkable, though, is the rapid, ingenious style of narration. From the opening —the close-up of a man's pale hand sliding down the bannister rail as he slips quietly out of a dark house—the camera seizes on the significant details which convey the narrative point of the scene. The result is a compression which gives the film continuous excitement.

For this compression, some credit is evidently due to Ivor Montagu, who was called in by the distributors when they found themselves dissatisfied with the first copy of the film. After specifying certain retakes, which Hitchcock shot, Montagu reedited the film and produced a version which the distributors accepted with delight. In view of later developments, however, there is no mistaking Hitchcock's primary responsibility for *The Lodger,* and for the ingenuity of its style in particular: A series of rapidly superimposed close-ups show alarm spreading as a new murder is reported; as the Buntings listen suspiciously to their lodger walking up and down in his room above them, we see a shot of the ceiling with his feet superimposed, walking to and fro, as though the floor were made of glass.

This inventiveness and visual dexterity was to form the basis of Hitchcock's style; they are the characteristics of a born storyteller, of one who delights to surprise and confound expectation, to build up suspense to a climax of violence and excitement. Strangely enough, though, the success of *The Lodger* did not lead Hitchcock to concentrate on this kind of film. He followed it with a return to romance, *Downhill,* which again starred Ivor Novello, as a noble boy who takes the blame for a chum's offense, is expelled from school ("Does this mean, sir, that I shall not be able to play for the Old Boys?"), and progresses downhill to the docks of Marseilles. There are interesting patches of technique: a delirium sequence as the hero is carried home on a cargo boat—scenes from his past superimposing and dissolving over shots of a gramophone playing in his cabin, the ships' engines turning over, the whole a powerful visual equivalent of discordant sound; and a daring subjective sequence as he lurches through the streets on his way home, the camera tracking and panning unsteadily to recreate his feverish impressions.

Three years, and six pictures, passed before, with *Blackmail* (1929), Hitchcock was able to find a story which suited as happily as *The Lodger*; in between there came a version of Noel Coward's *Easy Virtue* (which must have been almost as prodigious an achievement as Lubitsch's silent *Lady Windermere's Fan*), a boxing melodrama, a version of *The Farmer's Wife* and a couple of novelettes. Then at last Hitchcock hit on Charles Bennett's play, prepared a screenplay of it in collaboration with the author and Benn Levy, and shot it as a silent film. It was released, however, as Britain's first sound film, in part reshot and in part dubbed; it is thus of double interest—both for Hitchcock's uninhibited ingenuity in dealing with a new medium, and as a second example of his primary excellence in melodrama.

Blackmail is not as satisfactory as *The Lodger*; in construction it is less concise, less inevitable in progression. The connection of the first reel (the police at work) with the rest of the film is not well established; the scene in the artist's studio, in which Cyril Ritchard sings sub-Coward songs and attempts to seduce Anny Ondra is ludicrous in writing, setting, and handling; the famous chase, ending up with the blackmailer's fall through the dome of the British Museum, is too obviously tacked on to provide a spectacular climax. Also the film is weakened by the happy ending Hitchcock was forced (not for the last time) to substitute for the ironic fade-out he had planned.

Much of *Blackmail*, though, is excellent and survives in its own right. The everyday locales—a Corner-House restaurant, the police station, the little tobacconist's shop where the heroine lives with her parents, empty London streets at dawn—are authentic; the characters are believable; and at least one scene, between the blackmailer, the girl, and the detective, in which the detective does not know the guilt of the girl, the girl is too frightened to confess to it, and the blackmailer tries to play on the nerves of each, is worked up to a most successful tension. As in *The Lodger*, Hitchcock develops his story with a succession of felicitous, striking, or reveal-ing touches, particularly remarkable in this instance for the in-genuity with which they exploit the new dimension of sound. The portrait in the artist's studio, for instance, of a malevolently smiling jester is used as a sort of dumb commentator on the story—the last shot shows the picture carried away down the passage of the police station while the walls reecho to the sound of ironic laughter. Sound is used throughout with extraordinary freedom, for the period: to support continuity, as where the heroine, wandering in the streets after knifing her seducer, sees a man lying in a doorway, his hand dangling like the dead artist's; she opens her mouth to scream, and we cut to the scream of the landlady discovering the body of the murdered man. Two famous, and very effective, examples of the distortion of sound to convey a subjective impression of tension and near-hysteria occur as the girl sits miserably over breakfast the next morning. A garrulous neighbor is discussing the news: "I don't hold with knives. . . . No, knives isn't right . . . now, mind you, a knife's a difficult thing. . . ." Gradually all other words are mixed together in a monotonous blur, the word "knife" alone stabbing clearly out of the sound track over a close-up of the girl. "Cut us a bit of bread," says her father. The camera tilts down to a close-up of the knife; the girl's hand reaches out. Suddenly "KNIFE!" screams the voice, the hand jerks sharply up, and the breadknife flies into the corner of the room. A similar use of distortion and sudden crescendo conveys the girl's alarm at the sudden ringing of the shop bell: Instead of dying swiftly away, the sound of the bell is held for some four seconds, swelling up to a startling intensity.

Again like *The Lodger*, the films which followed *Blackmail* presented in the main a series of disappointments. *Juno and the Paycock* is straightforward filmed theater, well and respectfully handled; it is memorable however not so much for Hitchcock's contribution as for its perpetuation of some fine performances—in particular Sara Allgood's Juno, a figure that one sets beside Jane

Darwell's Ma Joad for its grandeur and humanity. *Murder,* which followed it, is an odd mixture, with some effective sequences—a midnight murder in a sleepy village, an exciting climax in a circus tent, with the murderer (Esmé Percy as an epicene trapeze artist) hanging himself from the big top. Amongst the enterprising uses of sound are one of the first uses of an overlaid track representing the thoughts running through a character's head while he shaves, and a not altogether successful experiment in expressionism—an impatient jury chanting in chorus against its one dissenting member. Long stretches of the film, though, are theatrical in the extreme, clogged with dialogue and dominated by an excessively stagey performance by Herbert Marshall.

None of Hitchcock's remaining films for British International (the producers of *Blackmail*) achieved much success. In 1933 he left to direct an unhappy excursion into musical comedy, *Waltzes from Vienna.* His career seemed to have reached its nadir when, with his infinite capacity for surprise, he rejoined Balcon at Gaumont British, renewed his association with Ivor Montagu (associate producer) and Charles Bennett (scriptwriter), and directed in a row a series of films which were to mark his most memorable and enjoyable contribution to the cinema.

The team of Hitchcock, Bennett, and Montagu remained in collaboration for three years, during which, with Balcon as producer, it was responsible for *The Man Who Knew Too Much, The Thirty-Nine Steps, The Secret Agent,* and *Sabotage.* In 1937 Balcon and Montagu left Gaumont British, but Bennett remained to write *Young and Innocent:* In 1938 Hitchcock made his last good British film, *The Lady Vanishes,* from a script by Launder and Gilliat. All these films are melodramas—stories of violence and adventure in which the emphasis is on incident rather than on characters or ideas. Hitchcock had himself come to realize that this was the form ideally suited to his talent and his temperament. In his autobiography, Esmond Knight quotes an illuminating *cri de coeur* on the set of *Waltzes from Vienna*: "I hate this sort of stuff," groaned Hitchcock, "melodrama is the only thing I can do."

Melodrama does not, of course, preclude common sense; with the exception of *The Lady Vanishes,* with its Ruritanian locale and its deliberate light comedy accent, these films gain a particular excitement from their concern with ordinary people (or ordinary-looking people) who are plunged into extraordinary happenings in the most ordinary places. This gives them immense conviction, and enables Hitchcock to exploit to the utmost his flair for the dramatic value of contrast. Instead of dressing up the Temple of Sun Worshippers —which covers the headquarters of the gang in *The Man Who Knew Too Much*—he presents it as a drab little nonconformist chapel,

bare and chilly, with a typically shabby congregation of elderly
eccentrics. In *The Thirty-Nine Steps,* the head of the organization
lives in a solidly respectable country house, and entertains the
[gentry of the] County to cocktails after Sunday morning service.
Verloc, the secret agent of *Sabotage,* runs an unpretentious subur-
ban cinema. The pursuit in *Young and Innocent* winds up at a *thé
dansant* at a seaside hotel. Similarly the people are conceived in
common-sense, unglamorized terms; the leading players (one hardly
thinks of them as stars) dress with credible lack of extravagance, get
dirty, behave like average human beings—neither brilliant nor
foolishly muddled. And supporting them are a multitude of equally
authentic minor characters, maids, policemen, shopkeepers, and
commercial travelers. This overall realism makes it all the more
thrilling when the unexpected occurs—as it inevitably does: Pretty
maids lie to the police without blinking an eyelid, harmless old
bird-fanciers are revealed as sabotage agents, old ladies who are
playing the harmonium one minute are whipping little revolvers
from their handbags the next.

The plots of these films are less important for themselves than
for the way they are unfolded. They are all stories of violence and
suspense, five exploiting in one way or another the excitements of
espionage and political assassination (of these, *The Thirty-Nine
Steps* and *The Man Who Knew Too Much* are perhaps the most
completely successful and continuously exciting), the sixth (*Young
and Innocent*) centering on the pursuit of a murderer by the young
man accused of his crime. In most of them the tensions of mystery
and intrigue erupt in a climax of public violence: The agents in
The Man Who Knew Too Much are exterminated in a street battle
which recalls the historic battle of Sidney Street; *The Thirty-Nine
Steps* winds up with shooting during a Crazy Gang show at the
Palladium; *The Secret Agent* has a train crash, *Sabotage* a time-
bomb exploding in a crowded bus, and *The Lady Vanishes* another
gunfight, between the agents of a foreign power and a party of
Britons stranded in a railway carriage in a central European forest.

These set pieces are not, however, isolated delights; the films are
continuously enjoyable for the brilliance and consistency of their
narration—a technique which shows the value of experience with
the silent cinema and the necessity of unfolding a story in visual
terms. Hitchcock has freely acknowledged his debt to Griffith; his
own style, at its best, has always been firmly based on cutting. In a
famous article on his methods of direction, published in *Footnotes
to the Film,* he states his *credo* specifically: "What I like to do always
is to photograph just the little bits of a scene that I really need for
building up a visual sequence. I want to put my film together on
the screen, not simply to photograph something that has been put

together already in the form of a long piece of stage acting. . . ." [1]
Besides being an admirable instrument for the building up of tension within the scene, Hitchcock's cutting contributes to the bold-
ness and ingenuity with which his plots are developed, with continuous speed and surprise. (His scripts are preplanned, his films
edited in the camera rather than the cutting room). We are precipitated at once into the middle of events—*Young and Innocent,*
for instance, starts brilliantly, at the climax of a murderous quarrel.
With a few happy strokes a locale is sketched in, an atmosphere
established; the stories proceed with a succession of ingenious visual,
or sound-and-visual, effects (the Hitchcock touch) as the celebrated
continuity from *The Thirty-Nine Steps* [the chambermaid discovers
the body and screams, but we hear the screech of a train emerging
from a tunnel in the next shot]; or the ominously sustained organ
note in *The Secret Agent* (a film packed with ingenious touches,
and Hitchcock's favorite of the series), which announces the death
of the Allied agent, strangled in the lonely little Swiss church.

Hitchcock's best films are in many ways very English, in their
humor, lack of sentimentality, their avoidance of the grandiose and
the elaborately fake. And these qualities were threatened when,
in 1939, he succumbed to temptation and signed a contract to work
in Hollywood for David Selznick. He was ambitious to make films
for the vast international audience which only Hollywood could
tap; also no doubt he was eager to work with the technical facilities
which only Hollywood studios could provide. It was particularly unfortunate, however, that Hitchcock chose the producer he did; for
Selznick is a producer who has always relied on pretentiousness, the
huge gesture, the imposing façade, to win success (*Gone with the
Wind, Since You Went Away, Duel in the Sun*). Almost in advance
Hitchcock was committed to all that is worst in Hollywood—to
size for its own sake (his first picture for Selznick was 2,000 feet
longer than any he had directed previously), to the star system for
its own sake, to glossy photography, high-toned settings, lushly
hypnotic musical scores.

The negotiations with Selznick were carried on while Hitchcock
was working on his last British film, *Jamaica Inn,* a dully boisterous
smuggling adventure with Charles Laughton. It was curious and
unhappily prophetic that his first film in Hollywood should also be
an adaptation from a Daphne du Maurier bestseller, *Rebecca*—a
less boring book, but equally Boots Library in its level of appeal.
Rebecca is a very skillful and competently acted film: Numerous
imitations employing the same theatricalities of suspense—the great
house dominated by a mysterious figure, the frightened girl, the

---

[1] [The article is reprinted in this volume, pp. 32–39.]

sinister housekeeper—emphasize the smooth plausibility of Hitchcock's handling. But the film as a whole is not recognizable as the work of the Hitchcock of, say, *The Thirty-Nine Steps*; it is at once bigger and less considerable.

The films which followed it in the next four years are of uneven quality, and represent no progression, no real acclimatization. *Suspicion* (the next-but-two) was an attempt to reproduce the high-class tension of *Rebecca,* again with Joan Fontaine; it succeeds only in ruining a fine thriller by Francis Iles, the story of a sensitive, unattractive girl married and murdered for her money by a handsome wastrel. By dressing her hair with severity and intermittently fondling a pair of horn-rims and a book on child psychology, Miss Fontaine effected the conventional compromise between glamor and realism successfully enough to win an Academy Award; but the film lacks excitement or conviction. The English backgrounds (Hunting, Church) are pure Burbank; and the ludicrous happy ending—neither written by Iles nor desired by Hitchcock—sets the seal of failure on the film.

Suspicion was preceded by a comedy, *Mr. and Mrs. Smith* (of which one would welcome a revival), and a thriller, *Foreign Correspondent*; after it came another reminiscence of the Gaumont British period, *Saboteur*. The earlier of these, written by Charles Bennett in collaboration with Joan Harrison, has excellent sequences embedded in a diffuse and vexatious story. The assassination of an elderly statesman in Amsterdam is brilliantly staged: rain drizzling, the square thronged with umbrellas, the news camera which fires a bullet, the assassin's escape through the crowd of bobbing umbrellas. There is a pleasantly sordid scene many reels later in which the kidnapped diplomat is grilled in a Charlotte Street garret, while a terrified German girl (in thick-lensed spectacles) sobs in terror by the wall; and the climax is worth waiting for—a transatlantic airliner shelled and nose-diving into the sea (entirely from within the plane), water crashing through the pilot's window, passengers fighting hysterically, and finally a handful of survivors clinging exhaustedly to a floating raft.

Saboteur is even more an affair of sequences, and is remarkable for its barefaced pilfering from almost every film Hitchcock had ever made. Its handcuffed hero and heroine (limp derivatives from *The Thirty-Nine Steps*) are pitched from one exotic location to another, individual episodes are directed with enjoyable virtuosity —the aircraft factory fire at the start, a gunfight in a cinema, the final megalomaniac climax on the Statue of Liberty—but the film as a whole has the overemphasis of parody.

It was not until 1943 that Hitchcock made a film which might be construed as an attempt—his last—to justify himself as a serious director. Before writing the screenplay of *Shadow of a Doubt,* he and Thornton Wilder went to live for two months in the little Californian town of Santa Rosa, where their story was to take place. Most of the film was shot there. As a result it has an everyday realism that is reminiscent of earlier days; and in its opening stages, a subtlety of characterization distinctly superior to them. Its central character is Charlie Newton, handsome and debonair, who lives by marrying and killing rich widows. Hard-pressed by the police, he comes to Santa Rosa to stay with his sister (who idolizes him) and her family: her quiet, respectable husband, her beautiful adolescent daughter, who feels that there is some special, secret bond between her and her uncle, and two smaller children. The film is at its best in its first half, establishing the family and their town, the impact of Uncle Charlie's arrival on each of them; experimenting once again with sound, Hitchcock adopted for these scenes a technique similar to Orson Welles in *The Magnificent Ambersons,* super-imposing one conversation over another, dovetailing, naturalistically blurring and distorting. The strange bond which seems to unite Young Charlie (the niece) with her uncle is subtly conveyed; the acting is excellent: Joseph Cotten as Charlie, bitter, arrogant, his smooth charm concealing a spirit wounded and festering, the exquisite Patricia Collinge, his sensitive, overstrung sister, Teresa Wright as Young Charlie, youthful and mercurial, waiting for love. In its later reels the film falls away; there is not the progression and development necessary to a serious study, and as a simple thriller (which is all perhaps Hitchcock would claim for it) it fails to sustain excitement and surprise. It remains, all the same, his best American film.

After *Shadow of a Doubt* Hitchcock completed one more picture in Hollywood, then ventured across the Atlantic to make his contribution to the Allied war effort. This came in the form of two short French-speaking films for the British Ministry of Information, *Adventure Malagache* and *Bon Voyage.* Each tells its story—the former of resistance activity in Vichy-dominated Madagascar, the latter of underground work in France—economically (most scenes are played in a single set-up), tastefully, and not very excitingly. A project for Hitchcock to direct a film about German concentration camps, for which he viewed a large quantity of documentary material, eventually came to nothing. This visit to Britain inspired no renaissance of style, no return to reality.

Almost, in fact, it appears to have precipitated his flight from it. From 1945 onwards the quality one associates with Hitchcock films is neither their excitement, nor their power to entertain, but their technical virtuosity. The trend had indeed already started in 1943, when he followed up *Shadow of a Doubt* with *Lifeboat.* For an hour and a half the camera remains in a lifeboat carrying eight survivors from a sunken Allied ship and one German, who turns out to be the captain of the U-boat which attacked it. The virtuosity of the direction is undeniable, and in a theatrical way the film is effective; but the attempts to build the story into a propagandist allegory, stressing the feebleness of a democracy in comparison with a dictatorship, were as unconvincing as (at this stage of the war) they were unnecessary. One remembers *Lifeboat* chiefly for its reintroduction of Tallulah Bankhead, and for some suspenseful episodes —a grim amputation carried out by the German with a clasp-knife, a realistically contrived storm.

Spellbound, with which Hitchcock returned to Selznick in 1945, also contains its entertaining passages of exhibitionism; its psychiatric background is futile and its Dali dream sequence merely pretentious, but one can enjoy the acid observation of the psychiatrists' common-room, and some facile patches of melodrama revolving around razors, glasses of milk, and the like.

It is unfortunate that even these are marred by a tendency to overplay, to inflate, a tendency which in *Notorious* swelled to an obsession and produced a film which shares with its successors, *The Paradine Case* and *Rope,* the distinction of being the worst of his career. In these films technique—lighting, ability to maneuver the camera in hitherto unimaginable ways, angles—ceases to be a means and becomes an end in itself; *Notorious* is full of large and boring close-ups. For hundreds of feet Ingrid Bergman and Cary Grant nuzzle each other in medium close-up, a sequence of embarrassing (because so thoroughly fake) intimacy. *The Paradine Case,* maimed from the start by Selznick's creaking script and a heavy roster of stars, is lit with magnificent but inexpressive artifice, contains further nuzzling by Ann Todd and Gregory Peck, and moves at a pace slower even than that of *Notorious.* And with *Rope,* a debilitated version of Patrick Hamilton's play, which abandons all the resources of cutting and lighting on the pretext of an experiment in technique, we come pretty well to a full stop.

Different though the results are, the experiment of *Rope* resembles the stylistic elephantiasis of *Notorious* and *Spellbound* in its preoccupation with technique, to the detriment of the material. The films, as a result, are neither good nor entertaining. To such highbrow accusations Hitchcock has a ready answer. To quote from an acutely critical article by Lawrence Kane (*Theater Arts,* May 1949): "*Spellbound* cost $1,700,000 and grossed $8,000,000. *Notorious* cost $2,000,000 and had enough love in it to take in $9,000,000. . . ." "Beyond that," said Alfred Hitchcock in a 1946 interview, "there's the constant pressure. You know—people asking, 'Do you want to reach only the audiences at the Little Carnegie or to have your

pictures play at the Music Hall?' So you compromise. You can't
avoid it. You do the commercial thing, but you try to do it without
lowering your standards. It isn't easy. Actually the commercial thing
is much harder to do than the other. . . ."

Disregarding the latter irrelevant (and untrue) argument, the
critic can only comment that Hitchcock's career in America has
suffered from more than compromise with commercialism (a compromise to which he has been no more exposed than any other
director of equivalent status). He is a director, in the first place,
who depends considerably on his scripts; in the last ten years he
has found no writer to give him what Bennett gave him at Gaumont
British. It is not a coincidence that his collaboration with Thornton
Wilder resulted in his best Hollywood film.

But *Shadow of a Doubt* hints at a more crippling limitation.
When Hitchcock left Britain, it was, at least in part, because he felt
that a chapter in his career was ended, and he was ripe for further
development. And in certain directions, it is true, Hollywood has
offered him vastly greater opportunities than Shepherd's Bush;
there are sequences in *Foreign Correspondent* and *Saboteur, Shadow
of a Doubt* and *Lifeboat* which outstrip anything in his earlier pictures for virtuosity and excitement. What these films lack is the
wholeness of their predecessors. The Gaumont British melodramas
succeed as works of art (however minor) because they attain a perfect, satisfying balance between content and style; the enlargement
which Hitchcock's style has undergone in Hollywood has been accompanied by no equivalent intensifying or deepening of sensibility
or subject matter.

Hitchcock has never been a "serious" director. His films are interesting neither for their ideas nor for their characters. None of the
early melodramas can be said to carry any sort of a "message"; when
one does appear, as in *Foreign Correspondent* and *Lifeboat,* it is
banal in the extreme—"You'll never conquer them," Albert Basser-
man wheezes on his bed of torture, "the little people who feed the
birds." In the same way, Hitchcock's characterization has never
achieved—or aimed at—anything more than a surface verisimilitude;
which, in a film where incident and narrative are what matters, is
perfectly proper.

The method, though, appropriate to *The Thirty-Nine Steps* and
The Lady Vanishes, is inappropriate to *Suspicion* and *Shadow of a
Doubt.* In these the more deliberate pace, the constant emphasis on
the players (dictated by the star system) directs our attention to the
characters; their emptiness becomes apparent, and the dramas fall
apart. *Suspicion* is not a failure simply because of its outrageous
volte face at the end; the characters have never begun to live, so
there is nothing really to destroy. In *Shadow of a Doubt* an at-

mosphere and a complex of relationships of some subtlety is established—only to dwindle conventionally instead of developing. *Notorious* presents an unpleasant but by no means uninteresting situation, which is thrown away largely because characterization is sacrificed to a succession of vulgar, superficial effects. In films like these, in *Rope* and *The Paradine Case*, even the old skin-deep truthfulness has been lost; Hitchcock's attitude towards his characters (as towards his audience) would seem to have hardened into one of settled contempt.

Hitchcock's progression from *The Pleasure Garden* to *Rope* is aesthetically pleasing; on the graph it would appear a well-proportioned parabola. But he does not oblige us by bringing his career to so satisfying geometric a close ("I am interested only in the future," he says). At the time of writing he has a period barnstormer, *Under Capricorn,* already completed, and a modern thriller, *Stage Fright,* almost finished. What is to be expected from these? Prophecy is always rash, but it is safe to assume that neither will present a dramatic reaction from the standards of showmanship which he has set himself in his "International" period; at their worst they will be heavy, tedious, glossed, at their best, ingenious, expert, synthetically entertaining. They will make a lot of money. Which, Hitchcock would reply, "is why they were made!" But at this point the wise critic resists the temptation to enter once again the vicious circle, and withdraws.

Hitchcock versus Hitchcock
by ANDRÉ BAZIN

I trust that the following account of my encounter with Alfred Hitchcock will not disappoint his wildest partisans. They may accuse me of being unworthy of the privilege of confirming all their insights. Certainly, where I am in doubt, I would prefer to give them the benefit of that doubt. I cannot say that the combined efforts of Schérer, Astruc, Rivette, and Truffaut have entirely convinced me of Alfred Hitchcock's flawless genius, particularly in his American work, but they have at least persuaded me to question my previous skepticism. Consequently, I can report that I approached my assignment in good faith and a constructive spirit by conscientiously assuming the point of view most favorable to the director and by insisting on his recognizing for himself and by himself every last morsel of meaning French critics had assigned to his films. Moreover, I would have been delighted if his answers had vindicated his champions and if the reservations I had formulated about such works as *Rope, The Paradine Case,* and *I Confess* had been reduced to rubble.

Before going any further, however, I propose some critical axioms, which the Hitchcockians may scorn as useless and undignified alibis.

I will begin with an embarrassingly personal anecdote. Once upon a time I analyzed a certain scene in *The Little Foxes,* the one in which Marshall is seen about to die on the stairway, in the background, while Bette Davis sits immobile in the foreground. The fixed gaze of the camera seemed to me to be intensified (moreover, if I remember correctly, the remark came from Denis Marion) by the fact that in the course of his movement the actor moved out of the line of vision and then came in again a little further away,

From Cahiers du Cinéma in English, *no. 2 (1966). Originally published as "Hitchcock contre Hitchcock" in* Cahiers du Cinéma, *no. 39 (October 1954). Reprinted by permission of Grove Press, Inc.*

while the lens—identified somehow with Bette Davis' implacability —did not deign to follow him.

When I attended the Brussels Festival in 1948 I had occasion to meet William Wyler, whose native language is French, and I explained my interpretation to him. Wyler seemed astonished. He insisted he had done everything quite simply with no intellectual premeditation. As for my crucial point about Marshall's departure from the field of vision, Wyler explained the specific reason: Marshall had a wooden leg and had difficulty climbing stairs; his eclipse permitted a double to be substituted for the last few seconds of the scene.

The anecdote was too funny to ignore even though the joke was on me. I reported it in "Le Film d'Ariane" (*Roman Holiday*) in *Ecran Français* under the collective signature of the Minotaur and held my ground as far as my initial analysis was concerned. My candor was rewarded with an ironic letter about critics of Spanish inns from some sly type reminding me of the Minotaur's note which I was forced to ignore in order to continue to ascribe to Wyler aesthetic calculations, a hypothesis that he himself had demolished.

I have verified the truth of this edifying story on several other occasions. There are, occasionally, good directors, like René Clément or Lattuada, who profess a precise aesthetic consciousness and accept a discussion on this level, but most of their colleagues react to aesthetic analysis with an attitude ranging from astonishment to irritation. Moreover, the astonishment is perfectly sincere and comprehensible. As for the irritation, this often springs from an instinctive resistance to the dismantling of a mechanism whose purpose is to create an illusion, and only mediocrities gain, in effect, from malfunctioning mechanisms. The director's irritation springs also from his resentment at being placed in a position that is foreign to him. Thus, I have seen a director as intelligent (and conscious) as Jean Grémillon play the village idiot and sabotage my debate on *Lumière d'été* evidently because he did not agree with me. And how can I say he is wrong? Is not this impasse reminiscent of Paul Valéry's leaving the lecture hall where Gustave Cohen had presented his famous commentary on *Cimetière Marin* with a word of ironic admiration for the professor's imagination? Must we conclude then that Paul Valéry is only an intuitive artist betrayed by a pedant's textual analysis and that *Cimetière Marin* is merely automatic writing?

As a matter of fact, this apparent contradiction between the critic and the author should not trouble us. It is in the natural order of things, both subjectively and objectively.

Subjectively, because artistic creation—even with the most intellectual temperaments—is essentially intuitive and practical: It is a matter of effects to attain and materials to conquer. Objectively, because a work of art escapes its creator and bypasses his conscious intentions, in direct proportion to its quality. The foundation of this objectivity also resides in the psychology of the creation to the extent—inappreciable—to which the artist does not really create but sets himself to crystallize, to order the sociological forces and the technical conditions into which he is thrust. This is particularly true of the American cinema in which you often find quasi-anonymous successes whose merit reflects, not on the director, but on the production system. But an objective criticism, methodically ignoring "intentions," is as applicable to the most personal work imaginable, like a poem or a picture, for example.

This does not mean that knowing *auteurs* personally, or what they say about themselves and their work, may not clarify the critic's conception, and this is proven by recently taped interviews that we have published. These confidences, on the contrary, are infinitely precious, but they are not on the same plane as the criticism I am discussing; or, if you will, they constitute a pre-critical, unrefined documentation, and the critic still retains the liberty of interpretation. Thus, when Wyler told me he had had Marshall leave the field of vision merely in order to substitute a double, I thought to myself that the flaws in the marble were useful only to good sculptors and that it was of little importance that the camera's fixity was imagined to come out of a technical contingency. But the following day when I saw Wyler again, it was he who returned to the subject and explained to me that Marshall's going out of the frame was not part of his artistic intentions and that, in turn, the light and soft quality of the background (the stairway where Marshall is dying) had been asked of Gregg Toland in order to create an uneasy feeling in the spectator by the imprecision of the action's essential point. In this context, virtually the entire film was shot in deep focus. Quite possibly, the softness and the disappearance had the same function: to camouflage the substitution of the double for the actor. Simply in the case of the softness, the director was conscious of the effect and the means, which suffices to elevate material servitude to the dignity of artistic windfall. Unless, profoundly astonished that so many things could be seen in this unfortunate sequence, he dreamed it up during the night and, when he woke up the next morning, was retrospectively persuaded that he had done it on purpose. It is of no real importance in terms of Wyler's glory and the excellence of *The Little Foxes,* but I am more partial to this explanation than to my original interpretation.

I make the foregoing observations in order to reassure and encourage those who, in this same issue of *Cahiers du Cinéma* (No. 39, October 1954), will credit Alfred Hitchcock with more talent than is implied in this interview. I am also perfectly conscious of not having pushed the *auteur* of *The Lady Vanishes* to a point where I could get past his defenses. Also the relatively serious nature of my questions undoubtedly had little in common with what he was accustomed to in American interviews, and the sudden change in critical climate may have upset him. Besides, people say that his answers anywhere tend more to mask than to reveal, and his penchant for straight-faced jesting is familiar enough to lend credence to this interpretation.

But now that I have raised every possible objection against my interviews with Hitchcock, I might well add that I am personally convinced of my interlocutor's sincerity, and I do not suspect for a moment that he accommodated my questions in order that I might judge his work less severely.

We met the first time at the flower market in Nice. They were shooting a scuffle. Cary Grant was fighting with two or three ruffians and rolling on the ground under some pink flowers. I had been watching for a good hour, during which time Hitchcock did not have to intervene more than twice; settled in his armchair, he gave the impression of being prodigiously bored and of musing about something completely different. The assistants, however, were handling the scene, and Cary Grant himself was explaining Nice police judo techniques to his partners with admirable precision. The sequence was repeated three or four times in my presence before being judged satisfactory, after which they were to prepare to shoot the following sequence—an insert in close-up of Cary Grant's head under an avalanche of pink flowers. It was during this pause that Paul Feyder, French first assistant on the film, presented me to Hitchcock. Our conversation lasted fifty or sixty minutes (there were retakes) during which time Hitchcock did no more than throw one or two quick glances at what was going on. When I saw him finally get up and go over for an earnest talk with the star and the assistants, I assumed that here at last was a matter of some delicate adjustment of the *mise en scène;* a minute later he came towards me shaking his head, pointing to his wristwatch, and I thought he was trying to tell me that there was no longer enough light for color—the sun being quite low. But he quickly disabused me of that idea with a very British smile: "Oh! No, the light is excellent, but Mister Cary Grant's contract calls for stopping at six o'clock; it is six o'clock exactly, so we will retake this sequence tomorrow." In the course of that first interview I had time to pose nearly all of the questions I had had in mind, but the answers had been so disconcerting that, full of caution, I decided to use a counterinterrogation as a control for some of the most delicate points. Most gracefully, Hitchcock devoted another hour to me several days later in a quiet corner of the Carlton in Cannes. What follows comes out of these two interviews without, in general, any distinction between what was in the first or the second.

I must make it clear at the outset that I speak and understand English too poorly to manage without an interpreter. I had the good fortune to find in Sylvette Baudrot, French scriptgirl with the crew, more than a faithful dragoman, a persistent collaborator. I take this opportunity to thank her cordially.

I attacked in more or less these terms: "While traditional criticism often reproaches you for brilliant but gratuitous formalism, several young French critics, on the contrary, profess a nearly universal admiration for your work and discover, beyond the detective story, a constant and profound message. What do you think about that?"

Answer: "I am interested not so much in the stories I tell as in the means of telling them." There followed a long account of *Rear Window* in terms of the technical improvisations that gave the film its originality. The film takes us, once more, into the realm of the detective story. The investigation is conducted in this instance by a convalescing magazine photographer, obliged to stay in his room because his leg is in a cast. He is also to discover a crime and identify the criminal solely by observing the comings and goings through the windows of the apartment building across the courtyard. During the entire film the camera remains in the journalist's room and sees only what he can physically see, either with the naked eye or with the aid of binoculars, which in any event allows for the use of different lenses. Telescopic lenses were ultimately required to keep the action in some sort of meaningful dramatic focus. The construction of the set posed equally complicated problems in order to permit the protagonist to observe as much as possible of his neighbor's actions without falsifying the architecture of an American city. Hitchcock himself insisted that half the film's action should be silent because the journalist cannot be expected to hear his neighbors at the distance he sees them. Thus the director had to resort to the guile of "pure cinema" which he adored. In general, dialogue is a nuisance to him because it restricts cinematographic expression, and he reproaches several of his films for this restriction.

At this point I did not abandon the point of my original inquiry by taking up the fallacious opposition of form and content. What Hitchcock calls "means" may be, perhaps, only a more indirect (and more unconscious) manner of following, if not a subject, at least a theme. I insisted, therefore, on the unity of his work, and he agreed with me in a negative way. All he demands of a scenario is that it go his "way." Let us stick our foot in the crack of that door. What I wanted was the exact definition of this "way." Without hesitation, Hitchcock spoke of a certain relationship between drama and comedy. The only films that may be taken as "pure Hitchcock" (*sic*) are those in which he has been able to play with this discordant relationship. Although this is more a matter of the way of conceiving a story than content properly speaking, it is, all the same, no longer a question of simple formal problems. I risk the word "humor." Hitchcock accepts it; what he is trying to express may well be taken as a form of humor and he spontaneously cites *The Lady Vanishes* as conforming most closely to his ideal. Must we conclude from this that his English work is more "purely Hitchcock" than his American? Without a doubt, first of all because the Americans have much too positive a spirit to accept humor. He could never have made *The Lady Vanishes* in Hollywood; a simple reading of the scenario and the producer would have pointed out how unrealistic it would be to send a message with an old woman by train when it would be quicker and surer to send a telegram. He thought he would please his old Italian maid by taking her to see *The Bicycle Thief,* but all she felt was astonishment that the worker did not end up borrowing a bike: America is rapidly becoming less colorful. Moreover, in Hollywood films are made for women; it is toward their sentimental taste that scenarios are directed because it is they who account for the bulk of the box-office receipts. In England films are still made for men, but that is also why so many studios close down. The English cinema has excellent technicians, but English films are not "commercial" enough and Hitchcock declares, with pain mixed with shame, that they are idle there while he is working. But it is still essential for a film to bring in more than it costs; the director is responsible for other people's money, a great deal of money, and he has a duty, in spite of everything, to be commercial. Hitchcock told me that his "weakness" lies in being conscious of his responsibility for all this money.

What I am inserting here is parenthetical: At the time of our second interview, the question came up again. Hitchcock appeared to me to be somewhat conventionally concerned with correcting that indirect criticism of being commercial by affirming that it was easy to make an "artistic" film, but the real difficulty lay in making a good commercial film, a very feasible paradox, after all. Such as it was, the sense of his first self-criticism was unequivocal and the necessity of renouncing adult, masculine humor in order to satisfy American producers was presented as an exquisite torture. When he arrived from England, and saw the technicians standing in line with their mess-bowls, under the clock, at the door of Warners', he anxiously asked himself if, in all this hubbub, film could possibly still be concerned with a form of the fine arts.

Faithful to my role of Devil's advocate, I remarked that, in the Hollywood studios, perhaps he had gained a sumptuousness of technical means that just suited his inspiration. Had he not always been concerned with ingenious and sometimes complex technical effects in order to obtain certain effects of *mise en scène?* Categorical answer: The importance of the technical means placed at his disposal did not particularly interest him. To the extent that they rendered the film more costly they even augmented commercial servitude. To sum it up, his ideal is, under those conditions, to accomplish perfection of "the quality of imperfection." This rather oracular line was one of those about which I was determined to see Hitchcock again to pin him down to a more precise confirmation. My interpreter, Hitchcock, and I spent a good quarter of an hour on this one point. He maintained what he had said and commented on it, but it never became perfectly clear. His exact words, in English, were, "I try to achieve the quality of imperfection." I believe I understood that the quality in question was American technical perfection (lacking in the European cinema) and the "imperfection" that margin for liberty, imprecision, and, shall we say, humor that makes, for Hitchcock, the English cinéaste's position superior. Thus it is a question for the director of *I Confess* of achieving the almost impossible marriage of perfect technical execution through Hollywood's oiled and supple machinery with the creative stumbling block, the unforeseen Acts of God, as in the European cinema. I am paraphrasing here and forcing myself to give a résumé of a conversation that, as far as I can see, was persistently obscure due to my lack of intellectual agility with the English language but also, I strongly believe, due to Hitchcock's instinctive irony. For I noticed several times his taste for the elegant and ambiguous formulation that goes so far as to become a play on words. Chabrol became aware of this tendency several times in Paris when Hitchcock made theological jokes based on "God" and "Good." This linguistic playfulness assuredly corresponds to a cast of mind but undoubtedly it is also a certain form of intellectual camouflage. For all that, I did not have the impression that this preoccupation affected our dialogue more than marginally. In general, the answers were clear, firm, and categorical. The circumstances were rare when, whether to correct the excessiveness of an affirmation that was a little too scandalous or paradoxical, or the question was particularly embarrassing, he used this sort of critical

humor to rectify something or to pirouette his way out of a statement. The general sincerity of his answers and, I even dare to say,
up to a certain point, their naïveté (if I am not misjudging how
much is bravado and how much paradox) was indirectly proven to
me by his reaction to one of my arguments. Always pursuing my
initial purpose of having him recognize the existence and the seriousness of a moral theme in his work, I decided, in default of
obtaining an acknowledgment from him, to suggest one myself,
borrowing for this the perspicacity of the fanatic Hitchcockians.
Thus, I had him notice that one theme at least reappeared in his
major films that, because of its moral and intellectual level, surely
went beyond the scope of simple "suspense"—that of the identification of the weak with the strong, whether it be in the guise of
deliberate moral seduction, as in *Shadow of a Doubt* where the
phenomenon is underlined by the fact that the niece and the uncle
have the same name; whether, as in *Strangers on a Train,* an individual somehow steals the protagonist's mental crime, appro-
priates it for himself, commits it, and then comes to demand that
the same be done for him; whether, as in *I Confess,* this transfer
of personality finds a sort of theological confirmation in the sacra-
ment of penitence, the murderer considering more or less consciously that the confession not only binds the priest as witness but
somehow justifies his acceptance of the guilty role. The translation
of such a subtle argument was not very easy. Hitchcock listened
to it with attention and intensity. When he finally understood it I
saw him touched, for the first and only time in the interview, by
an unforeseen and unforeseeable idea. I had found the crack in
that humorous armor. He broke into a delighted smile and I could
follow his train of thought by the expressions on his face as he
reflected and discovered for himself with satisfaction the confirmation in the scenarios of *Rear Window* and *To Catch a Thief.* It
was the only incontrovertible point made by Hitchcock's enthusiasts, but if this theme really exists in his work he owes it to
them for having discovered it.

However, I did not keep the initiative very long and the self-
criticism that he pursued was rather severe. *I Confess,* for example,
was rejected for its lack of humor—the comedy was not in step with
the drama. He is no longer enchanted with the players. Anne Baxter is an excellent actress but her personality is not socially true
to life in relation to Canada. He would have liked Anita Bjork
whom he greatly admired in *Miss Julie,* but Hollywood had been
scared off by her extramarital entanglements. (Ingrid Bergman's
troubles still lingered in the studio's memory.)

In opposition to this I remarked that he had, however, held on
to this subject, taken from a little-known French boulevard play,

given to him four years earlier by Louis Verneuil ("sold," he cor-
rected). If he had not shot it before, he explained, it was only
because Warner's was afraid of censorship; there was thus nothing
mysterious about the delay. Good; but must we not assume that
the films he produced would have a hold on his heart?—Not at
all, notably *Under Capricorn,* which, in spite of its failure, had
been principally a commercial enterprise. All of his efforts to save
something from this film were in vain. Hitchcock complained that
because of her fame Ingrid Bergman was no longer tenable. "All
the same," I said, "the brilliant sequences, continuous in time, that
called for the utmost in technical experience, in *Rope. . . .*" "Let
us talk about that!" he interrupted. "These continuous scenes were
boring enough later during the montage; there was nothing to cut!"

However, coming back to *I Confess,* I obtained an important
concession. When I praised the extreme technical sobriety, the intensity in austerity, it was not in order to displease him. It is true
that he applied himself here and that the film finds favor in his
eyes for these formal reasons. In order to characterize this rigor of
mise en scène it would be necessary to employ an epithet from the
"clerical vocabulary." . . . I suggested "Jansenist."—"What is Jansenist?" Sylvette Baudrot explained to him that the Jansenists were
the enemies of the Jesuits. He found the coincidence very droll
for he had studied with Fathers and, for *I Confess,* had been obliged
to free himself from his education! I did not tell him I would have
thought him, nevertheless, a better student. At least in theology.

Which, then, at least among his American films, did he consider
to be the most exclusively commercial and the least worthy of
esteem?—*Spellbound* and *Notorious.* Those that found grace in
his eyes? *Shadow of a Doubt* and *Rear Window.*

We have already spoken of the last one. What, in particular,
does he like in the first?—the truth, the social and psychological
realism, in the framework, naturally, of that dramatic humor we
have already defined. He was able to avoid the concessions and
commercial "fantasies" that more or less debased his other American films.

The interview came to an end, not because my interlocutor had
the air of becoming impatient, but because I could see no way to
bring the debate back to the essential. I come now to the formal
and secondary questions: Is it true that he never looked through
the camera?—Exactly. This task is completely useless, since all the
framing has been planned and indicated in advance by little drawings that illustrated the cutting technique. At my request he immediately executed several. If I may, I will add a personal comment here: It seemed to me, as much from certain precise points
made in the conversation as from statements gathered from Hitch-

cock's collaborators, that he had a permanent notion of *mise en scène,* that of a tension in the interior of a sequence, a tension that one would not know how to reduce either to dramatic categories or plastic categories but which partakes of both at the same time. For him it is always a question of creating in the *mise en scène,* starting from the scenario, but mainly by the expressionism of the framing, the lighting or the relation of the characters to the decor, an essential instability of image. Each shot is thus, for him, like a menace or at least an anxious waiting. From German expressionism, to whose influence he admits having submitted in the studios in Munich, he undoubtedly learned a lesson, but he does not cheat the spectator. We need not be aware of a vagueness of impression in the peril in order to appreciate the dramatic anguish of Hitchcock's characters. It is not a question of a mysterious "atmosphere" out of which all the perils can come like a storm, but of a disequilibrium comparable to that of a heavy mass of steel beginning to slide down too sharp an incline, about which one could easily calculate the future acceleration. The *mise en scène* would then be the art of showing reality only in those moments when a plumb line dropped from the dramatic center of gravity is about to leave the supporting polygon, scorning the initial commotion as well as the final fracas of the fall. As for me, I see the key to Hitchcock's style, this style that is so indisputable that one recognizes at a glance the most banal still from one of his films, in the admirably determined quality of this disequilibrium.

One more question to get off my conscience, the answer to which is easy to predict: Does he use any improvisation on the set?—None at all; he had *To Catch a Thief* in his mind, complete, for two months. That is why I saw him so relaxed while "working." For the rest, he added with an amiable smile, lifting the siege, how would he have been able to devote a whole hour to me right in the middle of shooting if he had to think about his film at the same time?

It was a charming way to end our conversation.

Why We Should Take Hitchcock Seriously

by ROBIN WOOD

Why should we take Hitchcock seriously?

It is a pity the question has to be raised. If the cinema were truly regarded as an autonomous art, not as a mere adjunct of the novel or the drama—if we were able yet to *see* films instead of mentally reducing them to literature—it would be unnecessary. As things are, it seems impossible to start a book on Hitchcock without confronting it.

Hitchcock has expressed repeatedly his belief in "pure cinema"; to appreciate his films it is necessary that we grasp the nature of the medium. We are concerned here with much more than is normally meant by the word "technique." Let us look at an example.

In *Marnie* Mark Rutland (Sean Connery) tries to "persuade" Strutt (Martin Gabel) not to take action against Marnie for robbing him (the money has been returned). Strutt protests that the lenient attitude asked of him is "fashionable," but "just wait till *you've* been victimized." Then we see Marnie enter the house. She has just been through a terrible accident, has had to shoot her beloved horse, still carries the pistol in her hand; she is distraught, and quite bereft of free will or the power of rational consideration. Reduce this to literary "content" detached from the images (from the movement of the images, and the movement from image to image) and there isn't much there: a straightforward dramatic situation with a rather dislikeable man making things unpleasant for the poor heroine. But compare this account with the force of the images on the screen. The cut from Strutt to Marnie gives us,

From the introduction to Hitchcock's Films *by Robin Wood (London: The Tantivy Press, in association with A. Zwemmer Ltd., 1965; Cranbury, N.J.: A. S. Barnes & Co., 1969), pp. 7–26. Reprinted by permission of The Tantivy Press. Title supplied.*

first, an ironic comment on Strutt's word "victimized." We see Strutt, complacent, self-righteous, a man devoid of human warmth and generosity taking his stand on a rigid, standardized morality; then we see Marnie, haggard, drawn, face drained of all color, clinging to a gun without realizing she is holding it, moving forward like a sleepwalker, "victimized" indeed by a whole complex past. It defines for us our moral attitude to Strutt's moral attitude, and is a comment on Strutt himself and all he stands for (he stands, in the film, for quite a lot: a whole attitude to life, a whole social milieu). There is also color: Mark wears a brown-flecked tweed suit and a brown tie, so that he blends naturally with the "natural" colors—predominantly browns and greens—of the decor of the Rutland house; the hard blue of Strutt's tie clashes with the green of the sofa on which he is sitting, making him out of place in the decor as his attitude to life is out of place in the Rutland milieu. The point, like so much that is important in the film, is *felt* rather than registered consciously by the spectator.

A novelist could give us some kind of equivalent for all this, could make us react along the same general lines; but he couldn't make us react in this direct, immediate way, as image succeeds image—he couldn't control our reactions so precisely in time. He could describe the way Strutt was sitting, but he couldn't *show* us Strutt sitting in precisely that way; we would all imagine it differently. He couldn't place us at a certain distance from Strutt to watch his awkwardness, his gestures, his changing expressions from a slightly low angle, so that we are aware at once of his absurdity and of his power. He could describe Marnie's appearance and analyze explicitly her emotional state from within (as Hitchcock can't); he couldn't show us her face, her way of moving; he couldn't place us above her, so that we look down on her with compassion; he couldn't place us in front of her and make us move with her, at her speed, so that we are caught up in her rhythm of moving, so that we share for a moment her trancelike state. The novelist can analyze and explain; Hitchcock can make us experience directly.

I have purposely chosen here a quite unremarkable (in its context) example which contains nothing obviously "cinematic"; it could be paralleled from almost any sequence in *Marnie*. It seems to me a fair representative specimen of that local realization that one finds everywhere in recent Hitchcock films, realization of theme in terms of "pure cinema" which makes the audience not only see but *experience* (experience rather than intellectually analyze) the manifestation of that theme at that particular point.

Or consider the moment in *Psycho* when Norman Bates carries his mother down to the fruit cellar. In literary terms there is almost nothing there: a young man carrying a limp body out of a room and down some stairs. Yet in the film the overhead shot with its complicated camera movement communicates to us precisely that sense of metaphysical vertigo that Hitchcock's subject requires at that moment—a sense of sinking into a quicksand of uncertainties, or into a bottomless pit—communicates it by placing us in a certain position in relation to the action and controlling our movements in relation to the movements of the actors. The cinema has its own methods and its own scope. We must beware of missing the significance of a shot or a sequence by applying to it assumptions brought from our experience of other arts.

The cinema—especially the Hollywood cinema—is a commercial medium. Hitchcock's films are—usually—popular; indeed, some of his best films *(Rear Window, Psycho)* are among his *most* popular. From this arises a widespread assumption that, however "clever," "technically brilliant," "amusing," "gripping," etc., they may be, they can't be taken seriously as we take, say, the films of Bergman or Antonioni seriously. They *must* be, if not absolutely bad, at least fatally flawed from a serious standpoint. And it is easy enough for those who take this line to point to all manner of "concessions to the box office," fatal compromises with a debased popular taste: Hitchcock returns repeatedly to the suspense thriller for his material; he generally uses established stars who are "personalities" first and actors second; there is a strong element of humor in his work, "gags" and "comic relief" which effectively undermine any pretensions to sustained seriousness of tone. To one whose training has been primarily literary, these objections have a decidedly familiar ring. One recalls that "commercial"—and at the time intellectually disreputable—medium the Elizabethan drama; one thinks of those editors who have wished to remove the Porter scene from *Macbeth* because its tone of bawdy comedy is incompatible with the tragic atmosphere; of Dr. Johnson's complaints about Shakespeare's fondness for "quibbles" and conceits; of Robert Bridges' deploring of the bawdy scenes in such plays as *Measure for Measure.* The argument, in all these cases, was basically much the same: Shakespeare allowed himself these regrettable lapses from high seriousness merely to please the "groundlings"; or, if we can't bear such a thought, we can comfort ourselves with the reflection that perhaps they were interpolations by someone else.

Now, one does not want to deny Shakespeare his imperfections, or Hitchcock his; indeed, a strong objection to much current French exegesis of Hitchcock, as to so much current critical work on Shakespeare, is that the writers tend to start from the assumption that their hero can do no wrong, and quite fail to make necessary

discriminations between different works, or admit occasional failures of realization within works. Such treatment does much harm,
by erecting barriers between artist and audience that are very difficult to break down; one is not responding at first hand to a work
of art if one approaches it in a spirit of uncritical veneration. But
what one does not want either Shakespeare or Hitchcock deprived
of is precisely the richness their work derives from the sense of
living contact with a wide popular audience. To wish that Hitchcock's films were like those of Bergman or Antonioni is like wishing
that Shakespeare had been like Corneille (which is what his eight-
eenth-century critics *did* wish). This implies no disrespect to Corneille (nor to Bergman and Antonioni), who can offer us experiences
that Shakespeare cannot; it is meant to imply that Shakespeare can
offer us richer experiences; and that if we somehow removed all
trace of "popular" appeal from Shakespeare and Hitchcock, then
we would have lost Shakespeare and Hitchcock.

That there are important distinctions to be drawn between
Shakespeare's audience and Hitchcock's, and important resulting
distinctions between the oeuvres of the two artists, I would not of
course wish to deny. And Hollywood is not as conducive to great
art as Elizabethan London, for many reasons too obvious to need
recounting here. But it is a matter of approaching these alleged
concessions and compromises from the right end. I remember once
expressing great admiration for Howard Hawks' *Rio Bravo;* I was
promptly told that it couldn't be a very good film, because Hawks
had used a pop singer (Ricky Nelson) and then introduced a song
sequence quite arbitrarily to give him something to sing. It didn't
apparently occur to the person in question to consider the song
sequence on its own merits in its context in the film (in terms of
strict thematic unity it is easier to justify than the role of Autolycus
in *The Winter's Tale,* which no one nowadays wishes away, though
it represents a "concession to popular taste" that, one feels, delighted
Shakespeare's heart very much as the *Rio Bravo* song sequence delighted Hawks'). It seems clear that the relationship of a Hitchcock
or a Hawks to his art is much more like Shakespeare's than is that
of a Bergman or an Antonioni; the sense of communication on many
levels precludes that self-consciousness of the artist that besets the
arts today, and fosters true artistic impersonality. All one asks for
Hitchcock is that people *look* at his films, allow themselves to react
spontaneously, and consider their reaction; that, for example, instead of assuming that *Vertigo* is just a mystery thriller (in which
case it is a very botched job, with the solution divulged two-thirds
of the way through and the rest, one supposes, total boredom), they
look without preconceptions at the sequence of images that Hitch-

cock gives us, and consider their first-hand responses to those images. They will then be led, very swiftly by the straightest path, to the film's profound implications.

It is precisely this refusal to look, react, and consider that vitiates most British criticism of Hitchcock's films. The characteristic "establishment" line may be fairly represented by the article "The Figure in the Carpet" by Penelope Houston in the Autumn 1963 issue of *Sight and Sound,* the immediate stimulus being in this case the appearance in England of *The Birds.* The article is in fact so characteristic, not only of the line on Hitchcock, but of the *Sight and Sound* critical approach in general, that it is worth studying, as a representative document, in some detail, which I do not propose to do here. A few comments, however, seem in order. What is most striking in the article is its almost exclusively negative character. Miss Houston examines cursorily some of the more obvious and easily dismissible excesses both of the book on Hitchcock by Eric Rohmer and Claude Chabrol, and of Jean Douchet's articles in *Cahiers du Cinéma;* when one looks for some clear positive lead one finds almost nothing. Miss Houston appears to include herself in that "general agreement that (Hitchcock) is a master," yet nothing she finds to say remotely supports such a valuation. There is continual evasion of critical responsibility, a refusal to follow any line of enquiry rigorously to its conclusion. Of *Vertigo* (which Miss Houston seems now to like, though it is impossible to discover why or how much or in what way) we learn that it "hypnotizes. . . . The first half moves like a slow, underwater dream. The second half . . . has the hallucinatory quality of a nightmare. By the end of the film, the audience ought to be as mad as James Stewart appears to be." And that is all. If we ask for some effort to explain this "hallucinatory quality"—its method, its purpose, its moral implications (for good or ill, intentional or unintentional)—we get no answer from Miss Houston. One is left to assume that any film is vaguely acceptable if it has a "hallucinatory quality."

Most of Miss Houston's article seems to rest on two supports, both critically insufficient to say the least—what Hitchcock has said, and what she herself assumes to have been his intentions in this or that film—and never, as far as one can judge, on a detailed first-hand study of the films themselves. Her closing paragraph (on *The Birds*) is representative: ". . . If *The Birds* is really intended as a doomsday fantasy, one can only say it's a lamentably inadequate one. . . ." It is at least equally inadequate if "really intended" as Oedipus Rex, Winnie the Pooh, or a pair of kippers; unfortunately, what concerns (or should concern) the critic is not what a film is "really intended" to be, but what it actually *is.* Miss Houston's remark also implies a set notion of what a "doomsday fantasy" ought to be like; anyway, her inability to cope with a complex work like *The Birds* is sufficiently revealed in that desire to package it neatly and tie on a ready-to-hand cliché label. ". . . But why not try the birds as the Bomb; or as creatures from the subconscious; or start from the other end, with Tippi Hedren as a witch? . . ." Why not, indeed? Go ahead, Miss Houston! ". . . One could work up a pretty theory on any of these lines. . . ." (Except that a minute's consideration of the film would be enough to show that one couldn't) ". . . If only one could suppress a conviction that Hitchcock's intention . . ." (intention again!) . . . "was an altogether simpler one. He scared us in *Psycho* enough to make us think twice about stopping at any building looking remotely like the Bates motel. He tries it again in *The Birds,* but we will happily go on throwing bread to the seagulls, because the film can't for long enough at a time break through our barrier of disbelief. . . ." And one cannot, at this point, suppress a conviction about Miss Houston's conviction: that it existed before she saw *The Birds,* which was accordingly judged by the criterion: Is this frightening me as much (and in the same way?) as *Psycho* frightened me? ". . . And a director who has told us so often that his interest lies in the way of doing things, not in the moral of a story, invites us to take him at his own valuation. . . ." An invitation any serious critic ought, one would have supposed, to decline, politely murmuring, "Never trust the artist—trust the tale." The whole article—with its apparent assumption that a few odd quirks are enough to justify the description of a director as a master, and one needn't look deeper, needn't at all inquire what this or that film actually *does*—is typical of the dilettantism that vitiates so much British film criticism. One respects far more—while considering them mistaken—those who find Hitchcock's films morally repulsive; at least their attitude is based on the assumption that a work of art (or entertainment) has the power to affect us for good or ill, and that one needs to examine a work in detail and in depth (in Paddy Whannel's words, "teasing out the values embedded in the style") before one can offer any valuation of it. This bland disregard of the need for moral concern (I would say commitment, had the word not become so debased by oversimplification) in some form links up with the astonishing confusion (or absence?) of values revealed by the fact that the issue of *Sight and Sound* that contained Peter John Dyer's contemptuous review of *Marnie* (another interesting representative specimen) also awarded *Goldfinger* three stars in its *Guide to Current Films* ("Films of special interest to *Sight and Sound* readers are denoted by one, two, three, or four stars"—do *Sight and Sound* readers accept such blackening of their characters?), and printed Dyer's eulogy of Roger Corman's vulgar and pretentious Bergman plagiarism, *The Masque of the Red Death.*

It is a fact, I think, that the chief obstacle in the way of a serious appraisal of Hitchcock's work for many people is Hitchcock's own apparent attitude to it; and it seems worth insisting for a moment on the fundamental irrelevance of this. What an artist says about his own work need not necessarily carry any more weight than what anyone else says about it: Its value can only be assessed by the test to which one must subject all criticism or elucidation, the test of applying it to the art in question and asking oneself how much it contributes towards either understanding or evaluating it. The artist's own utterances are more likely to have an indirect relevance, by telling us something further about his personality and outlook. I used to find maddening Hitchcock's refusal to discuss his work with interviewers on any really serious level; I have come to admire it. It seems so much in keeping with the character of the films themselves that their creator should be such a delightfully modest and unassuming man who makes no claims for his art outside the evidence of the films. He leaves that to the commentators—it is their job, not his. The attitude can be illustrated with that splendid moment in the interview in *Movie* 6. Hitchcock has been asked to express an opinion on the thematic "broadening out" of his later films: "Well, I think it's natural tendency to be less superficial, that's Truffaut's opinion—he's been examining all these films. And he feels that the American period is much stronger than the English period. . . ." He is the least self-conscious of great artists; that delight in creation of which the films themselves speak sufficiently clearly is accompanied by no sense of artistic self-importance. One cannot help thinking again of Shakespeare, content apparently to leave supreme masterpieces to the mercy of producers and actors, not even considering (as far as we know) that perhaps some of them ought to be preserved in print. The delight in creation was its own reward, its own justification.

But when one turns from British to French criticism of Hitchcock one is not made altogether happier. Eric Rohmer and Claude Chabrol deserve our gratitude for their pioneer work: Their book on Hitchcock constitutes a very serious attempt to account for the resonances his films can evoke in the mind.[1] One admires its many brilliant perceptions, and the authors' interest in the moral qualities of Hitchcock's films. It leaves one unsatisfied, however—with the feeling that, if that is all Hitchcock offers, the authors overrate him, and a complementary feeling that, no, that is not all he offers. Their analyses—they set out to cover all Hitchcock's films, British and American, up to and including *The Wrong Man,* more than forty

---

[1] [Eric Rohmer and Claude Chabrol, *Hitchcock* (Paris, 1957). See the selection from this book on *The Wrong Man* in this volume, pp. 111–16.]

films in 150 pages of text—have the effect of depriving the films of
flesh and blood, reducing them to theoretical skeletons. And one is
always aware, behind the enterprise, of the authors' sense of the need
to make Hitchcock seem "respectable." Accordingly, they play down
the suspense element and the comedy, and strip each film down to
some bald intellectual postulate. The sort of thing I mean is suggested by the remarks Miss Houston quotes in her *Figure in the
Carpet*—the work immediately under consideration is *Strangers on
a Train*: "One must consider Hitchcock's work in exactly the same
way as that of some esoteric painter or poet. If the key to the system
is not always in the door, or if the very door itself is cunningly
camouflaged, this is no reason for exclaiming that there is nothing
inside." This seems to me most misleading. The meaning of a Hitchcock film is not a mysterious esoteric something cunningly concealed
beneath a camouflage of "entertainment"; it is there in the method,
in the progression from shot to shot. A Hitchcock film is an organ-
ism, with the whole implied in every detail and every detail related
to the whole. In Spenser's words,

> For of the soul the body form doth take;
> And soul is form and doth the body make.

If we can't find the "soul" of a work of art expressed in its body,
informing and giving life to every limb, then we may be pretty
sure it is not worth looking for.

The Rohmer-Chabrol book suffers also from a refusal to look at
the films empirically; the authors have decided in advance on their
thesis and the films have to be made somehow to fit it: They are
Catholic films. Hence the significance of *Rear Window* "is impossible to grasp without a precise reference to Christian dogma." Were
this true, the film would surely have to be accounted a failure; a
successful work of art must be self-sufficient, its significance arising
from the interaction of its parts. But the assertion is of course ridiculous, and it leads Chabrol and Rohmer to distort the film drastically. According to them, for instance, the last shot of the film shows
James Stewart and Grace Kelly "in the same state, exactly as if
nothing had happened"; yet at the start of the film Stewart was on
the point of breaking with Kelly, and at the end they are engaged to
be married; and his back is now turned to the window. Again,
according to Rohmer and Chabrol, *Rear Window* offers us a straightforward denunciation of curiosity. A strangely equivocal denunciation, surely, that has most people coming away from the cinema
peering fascinatedly through other people's windows. And the
curiosity has, after all, brought to light a peculiarly hideous murder
and saved a woman from suicide. Hitchcock's morality, with its

pervading sense of the inextricability of good and evil, is not so simple.

The articles of Jean Douchet, published in *Cahiers du Cinéma* under the title *La Troisième Clef d'Hitchcock,* seem to me, for all their interpretative excesses and, again, a tendency to reduce things to abstractions, more generally persuasive; especially the accounts, in the last instalment, of *Rear Window* and *Psycho* (*Cahiers du Cinéma* No. 113). Douchet is not in the least embarrassed by Hitchcock's popularity or by his suspense techniques; indeed, these become the starting point of his analyses. But, again, I feel he oversimplifies, through a fixed determination to pursue one line of approach. To see the flats across the courtyard in *Rear Window* as a sort of cinema screen on which James Stewart "wish-fulfills" his secret desires is splendidly illuminating; but by the end of the argument we seem to have forgotten the fact that the murderer is a real man who has murdered a real woman, deposited real limbs around the country, and buried a real head in a real flower bed. Douchet's insistence on black magic, initially interesting, seems eventually to limit the significance of Hitchcock's films rather seriously. His treatment of *Psycho,* from the point of view of Hitchcock's control and manipulation of audience reaction, seems to me—despite disagreements over detail—brilliant, but it covers only one aspect of a complex film. Nevertheless, Douchet's work—and the Rohmer-Chabrol book—are so far beyond anything the British "Establishment" has given us in intelligence and critical rigor that one feels ungrateful in advancing any criticism of them. Even their moments of lunacy seem more intelligent than the relentless triviality of "establishment" reasonableness. The worst harm done by using black magic or "precise reference to Christian dogma" to explain a Hitchcock film lies in the resulting suspicion that the films cannot stand up without such support.

But, it will be objected, I am not answering my opening question: What can one adduce, positively, once all the false preconceptions have been cleared away, to encourage the doubters to believe that Hitchcock deserves serious consideration as an artist? . . .

First . . . , one might point to the *unity* of Hitchcock's work, and the nature of that unity. I mean of course something much deeper than the fact that he frequently reverts to mystery thrillers for his material; I also mean something broader and more complex than the fact that certain themes—such as the celebrated "exchange of guilt"—turn up again and again, although that is a part of it. Not only in theme—in style, method, moral attitude, assumptions about the nature of life—Hitchcock's mature films reveal, on inspection, a consistent development, deepening and clarification. Now almost any body of work by a single person will reveal unity of some sort, not only the oeuvre of an Ian Fleming, but of an Agatha Christie and even an Enid Blyton. But this steady development and deepening seems to me the mark of an important artist—essentially, that which distinguishes the significant from the worthless. There is discernible throughout Hitchcock's career an acceleration of the process of development right up to the present day, when the rate is such that the critic can perhaps be forgiven if it sometimes takes him a little time to catch up.

But within this unity—and this is something which rarely receives the emphasis it deserves—another mark of Hitchcock's stature is the amazing *variety* of his work. No need to point to the obviously "different" films like *Under Capricorn;* consider merely Hitchcock's last five films, made within a period of seven years, *Vertigo, North by Northwest, Psycho, The Birds, Marnie.* There are plenty of points of contact: the use of identification techniques to restrict the spectator, for a time, to sharing the experiences of a single consciousness, in *Vertigo* and *Psycho;* the birds in *Psycho* and *The Birds;* the theme of the parent-child relationship in *Psycho* and *Marnie;* the "therapeutic" theme in all five films (common to almost all Hitchcock's mature work). But even more striking is the essentially *different* nature of each work, in tone, style, subject matter, method; though one is constantly aware that all five are manifestations of a single genius, there is no repetition; each film demands a different approach from the spectator.

The thematic material of Hitchcock's films is much richer than is commonly recognized. True, he never invents his own plots, but adapts the work of others; again, one cannot resist invoking Shakespeare. Hitchcock is no more limited by his sources than Shakespeare was by his. The process whereby Greene's romance *Pandosto* was transformed into the great poetic drama of *The Winter's Tale* is not unlike that whereby Boileau and Narcejac's *D'Entre les Morts* became *Vertigo;* there is the same kind of relationship. Shakespeare found it necessary to make no greater changes in Greene's plot than did Hitchcock in Boileau and Narcejac's; the transmutation takes place through the poetry in Shakespeare and the *mise en scène* in Hitchcock. Nor is this a matter of mere decoration: Shakespeare's poetry is not an *adornment* for Greene's plot, but a true medium, a means of absorbing that plot into an organic dramatic-poetic structure; precisely the same is true of Hitchcock's *mise en scène* in *Vertigo.* Naturally, this transmutation of material does not always take place so successfully, without leaving intractable elements: Shakespeare has his *Cymbeline,* Hitchcock his *Spellbound.* In such cases, the artist tends to return later to a richer, more organic treatment of material which is closely related but from which the intractable elements are absent; thus Shakespeare writes *The Winter's Tale,* and Hitchcock makes (though this somewhat exaggerates the relationship to the earlier work) *Marnie.*

The mystery-thriller element is, in fact, never central in Hitchcock's best films, which is not to deny its importance. We could put it this way: "Suspense" belongs more to the method of the films than to their themes (insofar as any distinction is possible, such distinctions applied to organic works being necessarily artificial). Look carefully at almost any recent Hitchcock film and you will see that its core, the axis around which it is constructed, is invariably a man-woman relationship; it is never a matter of some arbitrary "love interest," but of essential subject matter. This will be readily granted, one supposes, of *Notorious, Vertigo,* or *Marnie;* but it is equally true of *Rear Window* and *North by Northwest.* Of the obvious exceptions, *Psycho* derives most of its power from its sexual implications and overtones—from the impossibility, for Norman Bates, of a normal sexual relationship—and *Rope* much of its fascination from the equivocal relationship between the two murderers (the whole action can be seen as a working-out of suppressed homosexual tensions). Of other partial exceptions, *Strangers on a Train* would be more completely satisfying were its central love relationship more fully realized; and the failure of *I Confess* seems due in part to the fact that the protagonist is a priest—the most interesting sequences are those dealing with his past love affair.

It is true that one can find a profound theme underlying almost anything if one is predisposed to search it out sufficiently diligently; what distinguishes a work of art is that this theme should be seen, on reflection, to inform the whole—not only the "content" (if there is such a thing as distinct from treatment; for what is the content of a film but sounds and images, and where else can we look for its style?), but the method. Think seriously about *Vertigo* and *Psycho,* and you will find themes of profound and universal significance; think again, and you will find these themes expressed in the form and style of the film as much as in any extractable "content." The subject matter of Hitchcock's *Vertigo* (as distinct from Boileau and Narcejac's) is no longer a matter of mere mystery-thriller trickery; it has close affinities with, on the one hand, Mizoguchi's *Ugetsu Monogatari,* and on the other, Keats' *Lamia.* To adduce these generally respected works is not to try to render *Vertigo* respectable by means of them—there is no sleight-of-hand involved of the "A is like B so A is as good as B" variety. *Vertigo* needs no such dishonest apologia, having nothing to fear from comparison with either work (it seems to me, in maturity and depth of understanding as in formal perfection, decidedly superior to Keats' poem if not to Mizoguchi's film). I want merely to ensure that the reader considers Hitchcock's film, not the plot on which it is based; the merits and demerits of Boileau and Narcejac are as irrelevant to *Vertigo* as are those of Greene to *The Winter's Tale*.

More practically, perhaps, in answer to my opening question, one can point to the disturbing quality of so many Hitchcock films. It is one of the functions of art to disturb: to penetrate and undermine our complacencies and set notions, and bring about a consequent readjustment in our attitude to life. Many refer to this quality in Hitchcock but few try to account for it; how often has one heard that a certain film is "very clever" but "leaves a nasty taste in the mouth" (*Shadow of a Doubt, Rope, Strangers on a Train, Rear Window* . . .). This "nasty taste" phenomenon has, I believe, two main causes. One is Hitchcock's complex and disconcerting moral sense, in which good and evil are seen to be so interwoven as to be virtually inseparable, and which insists on the existence of evil impulses in all of us. The other is his ability to make us aware, perhaps not quite at a conscious level (it depends on the spectator), of the impurity of our own desires. The two usually operate, of course, in conjunction.

This disturbing quality is frequently associated with the Hitchcockian "suspense," and it is this which I would like to consider next. It is very rarely a simple thing, very rarely "mere" suspense; but it is not easy to define, since it has many functions and takes many forms. Jean Douchet's definition—"the suspension of a soul caught between two occult forces, Darkness and Light"—although one of those phrases that sounds better in French than in English, strikes me as too abstract and generalized to be much help. Starting with something concrete, let us attempt two obvious but illuminating exercises in "practical criticism."

Compare, first, the crop-dusting sequence in *North by Northwest* with the helicopter attack in *From Russia with Love* (there is a fairly clear relationship between the two).[2] The difference in quality will seem to some readers too great and too obvious for the comparison to be worth making, but its purpose is not to score easily off a bad film but to help us define the quality of the suspense in Hitchcock. It is worth, perhaps, pointing out that *From Russia with Love* represents precisely that pandering to a debased popular taste that Hitchcock is widely supposed to be guilty of; the most hostile commentator would find difficulty in paralleling its abuses of sex and violence in any Hitchcock film. The film itself, in fact, need scarcely detain us; it will be generally agreed that the sole *raison d'être* of the helicopter sequence is to provide a few easy thrills.

---

[2] [For Hitchcock on this scene, see this volume, pp. 22–23. See also the drawings and analysis on pp. 145–73.]

From a purely technical viewpoint (if such a thing exists) the
Hitchcock sequence is clearly incomparably superior: It is prepared
with so much more finesse, shot with so much more care, every shot
perfectly judged in relation to the build-up of the sequence. Delicacy
and precision are themselves strong positive qualities. In comparison,
the Bond sequence is messy and unorganized, the *mise en scène*
purely opportunistic. But there is far more in question here than
the ability to construct a "suspense" sequence; the suspense itself
in *North by Northwest* is of a different order. The suspense in the
Bond sequence is meaningless: The attack is just an attack, it has
no place in any significant development, there is no reason apart
from plot—no *thematic* reason—for it to happen to Bond then or
to happen in the way it does; it has no effect on his character. The
suspense consists solely of the question: Will he get killed or not?
and as (a) we know he won't and (b) there seems no possible reason
to care if he does, it has no effect beyond a purely physical titillation.
In *North by Northwest* the crop-dusting sequence has essential
relevance to the film's development. The complacent, self-confident
Cary Grant character is shown here exposed in open country, away
from the false security of office and cocktail bar, exposed to the
menacing and the unpredictable. The man who behaved earlier
as if nobody mattered except himself, is here reduced to running
for his life, scurrying for cover like a terrified rabbit; he is reminded
—and we, who found him smart and attractive in his accustomed
milieu, are reminded—of his personal insignificance in a vast,
potentially inimical universe. The sequence marks a crucial stage
in the evolution of the character and his relationships, and through
that, of the themes of the whole film. If the character were not
attractive, for all his shortcomings, our response would be merely
sadistic, we would delight in the spectacle of an unpleasant man
getting his deserts; but we have become sufficiently identified with
him for our suspense to be characterized by a tension between conflicting reactions to his predicament.

North by Northwest is not, however, one of the Hitchcock films
that evokes a really disturbing or complex response. The instance
I have given, however stunning, shows Hitchcock's suspense at its
simplest (in the context, that is, of his recent work). A second comparison will take us a stage further, and once again I shall choose
a film that bears a clear relationship to Hitchcock's work: Robert
Aldrich's *Whatever Happened to Baby Jane?*, made a few years
after the great box-office success of *Psycho*. The comparison is the
more interesting in that Aldrich's film has its defenders, and
Aldrich himself some intellectual pretensions. Consider the sequence in which Joan Crawford, the crippled sister, struggles down-
stairs to telephone for help while Bette Davis, who is victimizing,

perhaps killing her, is on her way back to the house from town. One is aware here primarily of a sense of great effort which is not entirely explicable in terms of an attempt to convey the agony of the crippled woman's rung-by-rung descent. Indeed, if almost nothing in the film works, it certainly is not for want of trying: Every incident is milked for every drop it can yield and more, so that one often becomes embarrassedly aware of Mr. Aldrich tugging determinedly at a dry udder. But what is the purpose of this particular suspense sequence? To arouse our pity for a helpless woman? But it is totally unnecessary to go to these lengths to do that, so we cannot help suspecting that we are rather being asked to relish her suffering. The suspense seems to me entirely gratuitous, in fact. It carries no implications beyond "Will she or won't she?"; no overtones or resonances. The cross-cutting between the sisters has no purpose beyond prolonging the agony. Since Bette Davis is unaware of what is happening, there is no complementary struggle on her side and the moment of her return is entirely fortuitous. In any case, the film's trick ending, in the manner of *Les Diaboliques,* makes nonsense of everything that has gone before.

Compare, first, the cross-cutting between the tennis match and the murderer's journey to deposit the incriminating lighter in *Strangers on a Train.*[3] The tension generated here has meaning in that it arises from a trial of skill and endurance for both men (the winning of the match, the regaining of the lighter from the drain); all manner of resonances are aroused in the mind (stopping just short, perhaps, of a fully formulated symbolism) by the cross-cutting between the hand straining down into darkness and the struggle for victory in the match in brilliant sunshine; and the most complex reactions are evoked in the spectator, who cannot help responding to the efforts of *both* men. Or compare with the Aldrich sequence the scene in *Rear Window* where Grace Kelly is surprised while searching the murderer's rooms. Our suspense here is inseparable from the suspense of James Stewart, who is responsible for her danger but quite powerless to *do* anything; it is as if *we* had sent her there and must now watch her pay for *our* curiosity. In other words, the suspense has that characteristic Hitchcockian *moral* quality, the experiencing of the suspense being an essential factor in the evolution of the James Stewart character and an integral part of its complex meaning.

Instances could be multiplied: the celebrated Albert Hall sequence from *The Man Who Knew Too Much* (1956 version), where the suspense is the outward projection of the agonizing conflict

---

[3] [For another analysis of this scene, see Ronald Christ's article in this volume, pp. 104–11.]

within the heroine's mind—Hitchcock's way of making us share
that conflict; the descent of the staircase at the end of *Notorious,*
where Ingrid Bergman is being rescued from much more than a
gang of spies and death by poisoning, and where our response is
further complicated by a certain compassion for the Claude Rains
character. Enough has perhaps been said to demonstrate the com-
plexity of this concept of "suspense" in Hitchcock's films. I said
earlier that it belonged more to his method than to his themes
(while denying that any clear-cut distinction between the two was
possible). It is sometimes his means of making the spectator share
the experiences of the characters; it sometimes arises from a tension
in the spectator between conflicting responses; it is sometimes not
entirely distinct from a growing discomfort as we are made aware
of our own involvement in desires and emotions that are the
reverse of admirable. It is one of the means whereby we *participate*
in Hitchcock's films rather than merely watch them; but this does
not constitute a definition. We must always bear in mind the complex moral implications of the experiences we share or which are
communicated to us.

In fact, of the many Hitchcock imitations I have seen, the only
one (Ricardo Freda's *Terror of Dr. Hitchcock* is in a different cate-
gory, being less an imitation than a *hommage,* with a highly personal character of its own) that catches something of the true quality of Hitchcockian suspense—and that only for a few moments—
is Stanley Donen's *Charade.* I am thinking of the sequence near
the end among the pillars outside the Comédie Française where
Audrey Hepburn has to choose, in a matter of seconds, between
two men who are both demanding her trust, and the suspense is
the projection of the conflict within her between instinct and reason—shall she trust the man she loves but whose behavior has been
extremely equivocal, or act rationally and give the treasure she is
carrying to the other man? These moments apart, the film is a
shallow pastiche.

The theme adumbrated here—the necessity for trust above all,
whatever the risks—is the theme of one of Hitchcock's early, and
not entirely satisfactory, Hollywood films, *Suspicion.* I pass to this
now because it offers a convenient focal point for disentangling
two threads which run through Hitchcock's later work and, while
they do not in themselves *explain* his films, offer a means of access
to them. I had better say that the mystery surrounding the genesis
of this film—whether or not Hitchcock reversed the ending at the
last moment, and whether, if he did, it was against his own wishes,
is irrelevant; we are concerned only with the finished work. Such
a last-minute decision might help to explain the slightly "un-
cooked" quality of much of the film; since its whole significance,

as we have it, depends on the ending as it stands, it is difficult to believe that Hitchcock was strongly opposed to it.

First, what I call the *therapeutic* theme, whereby a character is cured of some weakness or obsession by indulging it and living through the consequences: Joan Fontaine falls in love with and marries Cary Grant. He is soon revealed as a liar and she comes to suspect that he is a murderer—eventually, that he is trying to murder *her*. The suspicions poison their marriage, making any open communication between them impossible. Only when they are eventually forced into the open is the fallacy exposed and, in the film's very last shot, a new start made. Two shots are worth singling out. The first gives us the moment of crystallization of the suspicions. The couple are playing a word game with the husband's best friend; as the two men talk, the woman's hands finger the letters on the table, absently arranging them, and suddenly they have formed the word "Murder." Immediately she "realizes" that her husband is planning to kill the other man—a marvelous depiction of the way the conscious mind can be guided by the subconscious; in its context, of how deeply entrenched values can manipulate conscious thought. For Joan Fontaine, at the outset, is a dowdy, repressed young woman, a colonel's daughter, who has led a sheltered life characterized by the rigid values of respectability and a total ignorance of the outer world. She is irresistibly attracted to the man who represents glamour and reckless, carefree abandon; but he represents also a total rejection of everything her family background and upbringing has stood for; subconsciously, she *wants* him to be a murderer.

The second shot occurs later in the film, a long shot where we see Joan Fontaine, now certain that her suspicions are justified, standing in a black dress before a window whose framework casts around her a shadow as of a huge web. She is feeling herself the victim, the fly caught in the trap. But the image suddenly gives us the truth— she is in reality the spider, fattening herself on her suspicions in the center of the web she has herself spun; or she is both spider and fly at once, victim of her own trap. The image anticipates the even more powerful one in *Psycho* where Norman Bates, sitting in his room beneath stuffed birds of prey, becomes, simultaneously, the bird (from his resemblance to it) and its victim (from his position under it).

The second thread is the extension of this "therapy" to the spectator, by means of encouraging the audience to identify. The outlook of the Joan Fontaine character is a very common one, certainly not restricted to colonels' daughters. From the time of her marriage onwards, we are restricted to the one consciousness: We know only what she knows, see only what she sees; we share her suspicions and learn from experience with her. With her, we find the Cary Grant character attractive: He is so romantic and dashing, so careless of mundane cares and restraints. But with her, we are gradually dismayed by his excesses: The reckless abandon with other people's money—and other people's feelings—comes to appear very unpleasant. So we become ashamed of having found him so attractive; if he were a complete blackguard, now, we would be exonerated, merely the victims of deceit, and we would be revenged on him when his downfall came. As Joan Fontaine's fingers arrange those letters into the word "Murder," the camera places us in her position: They are *our* hands. The film endorses the man's attitude to life no more than the woman's: If the limitations of her inhibited, sheltered respectability are chastized, so is their inevitable complement—the attraction towards total irresponsibility. And always it is our own impulses that are involved. Not only the characters'.

Hitchcock

by ANDREW SARRIS

Alfred Hitchcock is the supreme technician of the American cinema. Even his many enemies cannot begrudge him that distinction. Like Ford, Hitchcock cuts in his mind, and not in the cutting room with five different setups for every scene. His is the only contemporary style that unites the divergent classical traditions of Murnau (camera movement) and Eisenstein (montage). (Welles, for example, owes more to Murnau, whereas Resnais is closer to Eisenstein.) Unfortunately, Hitchcock seldom receives the visual analysis he deserves in the learned Anglo-American periodicals devoted ostensibly to the art of the cinema. Pages and pages will be expended on Resnais's synchronized tracks in *Last Year at Marienbad,* but the subtler diminuendo of Hitchcock's cross-tracking in the American remake of *The Man Who Knew Too Much* will pass by unnoticed. Truffaut, Chabrol, and Resnais can pay homage to Hitchcock, but the Anglo-American admirers of Truffaut, Chabrol, and Resnais will continue to pass off Hitchcock as a Continental aberration. "The Master of Suspense" is thus virtually without honor in his own countries.

Hitchcock's art is full of paradoxes. *The Birds,* for example, reveals a rigorous morality coupled with a dark humor, but the theme of complacency that runs through all his work is now so explicit that it is generally misunderstood. Hitchcock requires a situation of normality, however dull it may seem on the surface, to emphasize the evil abnormality that lurks beneath the surface. Hitchcock understands, as his detractors do not, the crucial function of counterpoint in the cinema. You cannot commit a murder in a haunted house or dark alley, and make a meaningful statement to the audience. The spectators simply withdraw from these

From The American Cinema *by Andrew Sarris (New York: E. P. Dutton & Co., Inc., 1968), pp. 57–61. Copyright © 1968 by Andrew Sarris. Published in a paperback edition and reprinted by permission of the publisher.*

bizarre settings, and let the decor dictate the action. It is not Us up there on the screen, but some play actors trying to be sinister. However, when murder is committed in a gleamingly sanitary motel bathroom during a cleansing shower, the incursion of evil into our well-laundered existence becomes intolerable. We may laugh nervously or snort disgustedly, but we shall never be quite so complacent again. Hitchcock's repeated invasions of everyday life with the most outrageous melodramatic devices have shaken the foundations of the facile humanism that insists that people are good, and only systems evil, as if the systems themselves were not functions of human experience. Much of the sick, perverse, antihumanistic humor sweeping through America today is an inevitable reaction to the sickening sentimentality of totalitarianism masquerading as all-encompassing humanism. Hitchcock has never been accepted as part of this fashionable sickness, and his unfashionableness is all to his credit. He insists, almost intolerantly, upon a moral reckoning for his characters and for his audience. We can violate the Commandments at our own psychic peril, but we must pay the price in guilt at the end. Hitchcock can be devious, but he is never dishonest.

Hitchcock's reputation has suffered from the fact that he has given audiences more pleasure than is permissible for serious cinema. No one who is so entertaining could possibly seem profound to the intellectual puritans. Furthermore, did not Santayana once observe that complete understanding extinguishes pleasure? No matter. Hitchcock's art will always delight the specialist because so much of it is rendered with an air of casualness. The iron is encased in velvet, the irony is simplicity—simplicity, however, on so many levels that the total effect is vertiginously complex. Beneath the surface melodrama of every Hitchcock film is a lively comedy of manners. In this regard, *The Farmer's Wife* of 1928 displays Hitchcock's flair for satiric pantomime much as Dreyer's *Master of the House* reveals the Great Dane apart from the shadow of eternity. Hitchcock's style is alive to the expressive potentialities of the slightest encounter. His cutting is the means by which he contradicts what people say by what they do. In the beginning of his career, he was attracted to montage as a mental language. In *Murder* (1930) a sleuth-type character thinks of the hot meal he is missing at home in order to stay at a dingy hotel, and Hitchcock cuts to a spatially abstract shot of roast duckling purely as a mental expression of a gourmet's grief. Hitchcock quickly abandoned this experiment in favor of the intricate editing of objects and glances within a scene. His films abound with objects as visual correlatives —the missing finger in *The Thirty-Nine Steps,* the crashing cymbals (not symbols) in both versions (1935 and 1956) of *The Man Who*

Knew Too Much, the milk chocolates on the assembly line in *Secret Agent,* the knife and time bomb in *Sabotage,* the doctored drink in *The Lady Vanishes,* the twitching eye in *A Girl Was Young* [*Young and Innocent*], the monogrammed pillowcase in *Rebecca,* the reverse-sailing windmills in *Foreign Correspondent,* the conjugally crossed skis in *Mr. and Mrs. Smith,* the sinister glasses of milk in *Suspicion* and *Spellbound,* the magisterial Statue of Liberty in *Saboteur,* the incriminating ring in *Shadow of a Doubt,* the concealed compass in *Lifeboat,* the key to the winecellar in *Notorious,* the hypnotic portrait in *The Paradine Case,* the omnipresent trunk in *Rope,* the shrunken head in *Under Capricorn,* the bloodstained doll in *Stage Fright,* the incriminating cigarette lighter in *Strangers on a Train,* the falling bicycle in *I Confess,* the latchkeys in *To Catch a Thief,* the bothersome corpse in *The Trouble with Harry,* the hallucinatory coiffure in *Vertigo,* the crop duster in *North by Northwest,* the motel shower in *Psycho,* the besieged telephone booth in *The Birds,* the papier-mâché flames in *Torn Curtain.* Hitchcock's objects are never mere props of a basically theatrical mise-en-scène, but rather the very substance of his cinema. These objects embody the feelings and fears of characters as object and character interact with each other in dramas within dramas.

The late James Agee perceived the novelistic nuances of Hitchcock's visual storytelling in *Notorious,* but most American reviewers have failed to appreciate the Hitchcockian virtues of vividness and speed as artistic merits. Hitchcock's economy of expression can be compared favorably to that of any of his colleagues or imitators. There is, for example, a Hitchcockian touch in John Huston's direction of *Reflections in a Golden Eye.* Marlon Brando's Captain Penderton is following a private with whom he is obsessed. There is a sound of a car crash behind Brando. The private and his buddies all turn to look at the crash. Brando keeps staring at the private. The Hitchcockian equivalent of this sequence occurs in *Strangers on a Train* at a tennis match involving a troubled participant played by Farley Granger. The visual coup of the sequence is the familiar joke of spectators at a tennis match swiveling their heads back and forth to follow the action. All heads swivel but one—that of a psychopathic murderer played by Robert Walker. The differences of camera placement and editing between the Hitchcock sequence and the Huston sequence is the difference between visual directness and visual obliqueness. Hitchcock gives the audience the point immediately with a device designed for maximum vividness. Huston's effect is slower in making its point. Paradoxically, however, Hitchcock is more oblique psychologically than Huston. Hitchcock's characters are ostensibly obsessed by the issues of a contrived melodrama. The Walker character has proposed to the Granger character an exchange of murders so that the police would be left without a plausible motive for either murder. The Granger character has never taken the proposition seriously enough to reject it flatly, and he is horrified to discover that he is an accomplice to one murder and expected to keep his end of the bargain by committing another. Walker stalks Granger everywhere, most memorably at the tennis match. It is only under the surface of the melodrama that the darker humor of homosexual obsessiveness comes into play as an added layer of meaning. Huston's treatment of homosexuality is much closer to the surface. Indeed, the Brando characterization of Captain Penderton is less subtle and repressed than the Carson McCullers original out of her Gothic novel. Hitchcock is ultimately more cinematic; Huston more literary. Hitchcock operates on many levels, Huston only on one. The beauty of Hitchcock's style is a function of its speed and efficiency in operating a time mechanism. Huston's personality is not expressed so much through the medium itself as in a sour reaction to its emotional facility.

Hitchcock has worked with big stars from Nita Naldi to Julie Andrews and from Ivor Novello to Paul Newman, but he has generally managed to impart Hitchcockian humor to the most distinctive personalities. The ultra-Hitchcockian performances are those of James Stewart in *Rope, Rear Window, The Man Who Knew Too Much,* and *Vertigo* and Cary Grant in *Suspicion, Notorious, To Catch a Thief,* and *North by Northwest.* Stewart and Grant gave Hitchcock the means he could not have got from any other actors. In return, Hitchcock gave Stewart and Grant meanings they could not have got from any other director. Nonetheless Hitchcock has seldom been the favorite director of his players. Thespians traditionally prefer weaker wills and more adaptable visions of life.

1. Hitchcock's first film appearance in *Blackmail* (1929), also the first English talking picture. Photo courtesy Leo Braudy.

2. Cary Grant and Ingrid Bergman discover the "Mac-Guffin" in *Notorious*. Reprinted by permission of Alfred Hitchcock. Photo courtesy Leo Braudy.

3. The tower shot from *Vertigo,* one of Hitchcock's most famous trick shots, also underlines his obsession with heights. Copyright © 1958 by Alfred J. Hitchcock Productions, Inc. Reprinted by permission of Alfred Hitchcock. Photo courtesy Cinemabilia.

4. Hitchcock's characters are often voyeurs, as are his audiences. Here Thelma Ritter, Grace Kelly, and James Stewart are equipped for some looking across the courtyard in *Rear Window*. Reprinted by permission of Alfred Hitchcock. Photo courtesy The Museum of Modern Art Film Stills Archive.

5. The socially proper and the eccentric interlock in the Gaumont-British cycle of the 1930s. Here John Gielgud and Peter Lorre as an odd duo in Hitchcock's favorite of the series, *The Secret Agent*. Reprinted by permission of Alfred Hitchcock and The Rank Organisation. Photo courtesy The Museum of Modern Art Film Stills Archive.

6. Hitchcock directing *Under Capricorn*. Since the picture is complete in his head before filming, he intervenes infrequently. André Bazin noted that "seated in his armchair, he gave the impression of being prodigiously bored and musing about something completely different." Reprinted by permission of Alfred Hitchcock. Photo courtesy The Museum of Modern Art Film Stills Archive.

7. Bruno (Robert Walker) and Guy (Farley Granger), doubles in *Strangers on a Train*. "Here is the gist of the plot: the exchange of murders Guy and Bruno have discussed, the ambiguous exchange of guilt and identity. . . ." Copyright © 1951 by Warner Bros., Inc. Reprinted by permission of Alfred Hitchcock and Warner Bros., Inc. Photo courtesy The Museum of Modern Art Film Stills Archive.

8. Theater as a metaphor for life is especially strong in the English films of the 1930s. Here it's "curtains" for the villain of *The Thirty-Nine Steps,* which begins and ends in a music hall. Reprinted by permission of Alfred Hitchcock and The Rank Organisation. Photo courtesy The Museum of Modern Art Film Stills Archive.

9. In *The Wrong Man,* Hitchcock's attempt at semidocumentary, he humiliates "a man in that which is most noble, his gaze." Here Henry Fonda after his false arrest. Copyright © 1957 by Warner Bros., Inc. Reprinted by permission of Alfred Hitchcock and Warner Bros., Inc. Photo courtesy Cinemabilia.

10. Behind the American motel is the Gothic house and its secrets. Here Tony Perkins goes up to see "mother" in *Psycho*. Reprinted by permission of Alfred Hitchcock. Photo courtesy Leo Braudy.

11. Cary Grant flees a plane trying to shoot him down on the prairie in Hitchcock's classic sequence from *North by Northwest,* his chief example of "avoiding the cliché" in visual suspense. The full sequence is analyzed on pp. 145-73. Copyright © 1969 by Metro-Goldwyn-Mayer, Inc. Reprinted by permission of Alfred Hitchcock and Metro-Goldwyn-Mayer, Inc. Photo courtesy Leo Braudy.

The Strange Case of Alfred Hitchcock, Part Three

by RAYMOND DURGNAT

For Chabrol and Rohmer, Hitchcock's vision is impregnated by a Roman Catholicism the severity of whose morality evokes Jansenism.[1] This severity may, however, have reached Hitchcock by a different route. Jansenism is the result, in French Catholicism, of the Calvinist influence, and Hitchcock's sense of a blind, implacable, cruel, yet, somehow, just, providence may well have come to him via the influence, on the British middle classes, of that Puritanism which is the English version of Calvinism. It's true that this influence has been somewhat overlaid, in English nonconformism, though not so much in Scottish, by the anti-Calvinism of Methodism. But the Jansenist theory meets an equal objection in the fact that Hitchcock was educated by Jesuits, who, theologically, are at the opposite pole to Jansenists. So one could argue that Hitchcock's morality is very English (and American) Protestant in that he swings between an optimistic (post-Wesleyan) vision and a grimly punitive view of human depravity.

Predestination was another Calvinist speciality, and a keynote of the Hitchcock world is its sense of a manipulated order—almost of predestination, with Hitchcock, like a Calvinist providence, arranging his narrative machines. Calvin accepts God's responsibility for evil, adding that good and bad are defined purely by God's verdict, not by their inherent characteristics (since this would mean that good and evil are God's God, and God would not be God, being subservient to them). Since God's will determines the nature of good and evil, and is sovereign, it is not for us to protest against

From Films and Feeling (*April 1970*). *Copyright* © *1970 by Raymond Durgnat. Reprinted by permission of the author.*

[1] [Eric Rohmer and Claude Chabrol, *Hitchcock* (Paris: Presses Universitaires de France, 1957). See this volume, pp. 111–16.]

any of his acts, however evil they seem, nor to deny that evil is evil. In saying this, Calvin answers, heroically, drastically, and cruelly, a problem which Christian orthodoxy has never managed to solve (if God created everything, how did evil enter the world?). Thus the Calvinist God is a despot, beyond good and evil. Associated with this view is the Calvinist emphasis on predestination. Non-Calvinist orthodoxy allows for predestination: "God foreknew and sorrowed, that you would, alas, abuse his gift of free will, and sin and damn yourself into his eternal fire." The Calvinist speciality is double predestination: "God created you such that you would damn yourself, and cannot feel sorrow for his own action. But nonetheless, you, being you, damned yourself and deserve to be punished." This relentlessness takes on a certain slyness, like a practical joke. Non-Calvinist orthodoxy disturbs our complacency by warning us that many who seem saved are damned and many who seem damned are saved. The grimmer Calvinist line has it that many who seem saved are damned but that all who, because they sin, seem damned, *are.*

The serene, implacable, indifferent face of the Calvinist God is evoked in those impassive faces; the Egyptian God past which the detectives hound their prey in *Blackmail,* or the Statue of Liberty, that American deity, from which the Nazi hangs in *Saboteur,* or the idealistic American presidents carved in Mount Rushmore. Indeed, Hitchcock's films are Calvinistic machines, the suspense lies in their predestinatory operations. It's significant that (as Chandler ruefully remarked, after his collaboration with Hitchcock on *Strangers on a Train*) the situations and the settings determine the action, to which character is made to fit.[2] Character is passive to the overall action, to the machinery of dream which is like the machinery of predestination. Character is merely smoke figures of Hitchcock's, or God's, opium visions, visions of crime and punishment, of innocence in nightmare.

Calvinism, like other extreme, heroic, and paradoxical positions, is an unstable one, and Hitchcock's vision can't be explained in exclusively Calvinist (or Jansenist) terms. In the first place, of course, there is Hitchcock's immense admiration for the sly insolence, the cheerful badinage of his heroes. There is the regulation "happy end" (Hitchcock's regular capitulation to which does more to undermine his films than any other box-office requirement) and often his happy ends allow his heroes to get away with things while suffering rather less from the suspense than, say, the innocent hero of *Phantom Lady.* There is the everyday, common-sense morality to which Robin Wood refers. There is a certain view of happiness

---

[2] [For Chandler's remarks, see this volume, pp. 101–4.]

—a compound perhaps, of the everyday contentment celebrated in *Blackmail* and *Shadow of a Doubt,* and of the slightly queasy sensuality typified by the glossy kissing of *Notorious.* The Calvinist grimness is alleviated by certain conventional visions which Hitchcock borrows and adapts. His heroes tend to be matinee idols in a nightmare; the happy end has behind it, not the force of God or providence or fate, but the audience's preference for reassurance; flexible empiricism is in the air anyway; the images of happiness are borrowed from everyday mediocrity or from glossy magazine fiction.

There is a streak, even, of rebellion against God, although it takes a peculiar form. Hitchcock *becomes* God. His characters clamber, perilously, about *his* portly, enigmatic face, for his, and our, amusement. Hitchcock, like Buñuel's Crusoe, looks down at the scurrying ants, and recognizes himself in them. The audience is in a middle position: part immune, part God, part victim. Hitchcock strolls by—God, incognito, unrecognized, unseeing. Hitchcock is like a God playing as many practical jokes on human beings as he can—jokes in rather poor moral taste. "You will be severely punished for not being good enough. The world isn't good enough either. Ha ha." Hitchcock's bland face begins to look less like that of the conventional Christian God than that of the Calvinist God whose cruelty is immune from all plea, but it also reveals a dreamy, mischievous, human pleasure. It begins to look like that of Sade. It's as that face is allowed to emerge that Hitchcock becomes most challenging (his two Christian films are suspiciously solemn, as if the intention was there, but not the experience). For once, his publicity image is truer than much criticism. Slogans like "Stories they wouldn't let me do on TV" implies a wry cruelty, a saturnine whimsicality which he shares with us, secretly, as individuals, but which, alas, isn't for public consumption. A sense of moral mischief is shared with Polanski—albeit Polanski is a little too sophisticated, a little too playful. The real meaning of *Rosemary's Baby* is "2,000 years of Christianity led only to us, so why shouldn't the devil become son of man? Things will go on much as before, or possibly be a bit more startling, a bit more fun." Polanski is New Morality, Hitchcock is Old Morality. *Rope* might be meant for people who, like Polanski, irresponsibly play with moral ideas. Hitchcock, like Greene, believes in the diabolic as a bedrock reality. Greene hopes that somebody up there will manage a few minor and furtive miracles somehow and give us the strength to go on joylessly suffering. Hitchcock believes in a healthy fear. Don't tempt providence (or the box office), lie low, be humorous and afraid, prudent and safe. Sade counseled bravado, Hitchcock mediocrity, and a little bit of luck. The Hitchcock vision is an uneasy suspension between a Calvinist fear of God, and an identification with a Sadistic God.

Between the two come humor, stoicism, and resignation. The "basic sympathetic flow" is present, but, somehow, intellectually, mechanically, so. Hitchcock is, discreetly, above it, ironically masterminding it, much as the Calvinist's God masterminds a sexual instinct with which he is in a gruelingly sado-masochistic relationship. In Hitchcockian sexuality there is much teasing, much dissatisfaction, much tussling for dominance. Handcuffs pull your limp hands to the stockinged thighs of the girl who wants to hand you over to the police [in *The Thirty-Nine Steps*]. . . .

Hitchcock's images of happiness are boring, his ventures into idealism don't lack nerve but fall flat on their face. How phoney— as Gavin Millar argues[3]—is that outrageous cornpatch in *Saboteur* with the blind wise man spotting his guest's innocence by the sound of his footsteps. Powell's *Peeping Tom*, with its wise blind woman scenting her guest's guilt, is nearer to a poetic vision, and to the Hitchcock truth. Marnie's lacquered visuals betray the story's harrowing elements; her psychopathic cure lacks all conviction but that of fashion, and is unreally agreeable even by the standards of Frank Sinatra's kicking the habit in *The Man with the Golden Arm*. But if Hitchcock's sexuality has its unconventional and troubling streak, its variations from the norm are prudent, stylized and oddly innocent. The manipulating and teasing of passive males by slyly overwhelming females (of James Stewart by Grace Kelly in *Rear Window*, of Cary Grant by Eva Marie Saint in *North by Northwest*) are deft little variations on a standard American theme (there's more sado-masochistic purification in Preminger's film, when Sinatra gets Kim Novak to lock him in to suffer withdrawal symptoms). Those long oppressive kisses come from some sumptuous, mail-order catalogue of erotic behavior (for Hitchcock they exemplify the lovers' refusal to stop spooning for life's little details; they might equally betray the cerebral coldness which enables them to kiss and dial; the ambiguity is interesting, not profound). The hints of deviation are no more striking than those in glossy advertisements (e.g., a cosmetics advertisement showing a girl's high heel planted firmly on the knuckles of a man's hand, helplessly reaching for an automatic; it might be a close-up from a Hitchcock film). The undoubted impact which Hitchcock imparts to such trivia comes, perhaps, from an effect analogous to his switching between moral fundamentalism and something more sophisticated. Here he switches between the fashionable daydream which slyly begins going wrong, a dream turning into a nightmare. Hitchcock sexuality be-

[3] [Gavin Millar, "Hitchcock *versus* Truffaut," *Sight and Sound* (Spring 1970): 41–44.]

gins that change; only in *Rear Window, Vertigo,* and *Psycho* does it go beyond it.

Hitchcock remains, prudently, within the limits of the teasing game. If *Psycho* has successive identification figures (Janet Leigh, Vera Miles, Anthony Perkins) it is because it has no principal identification figure; it needs only a butt, a fall guy, that is to say, the audience, which thinks it understands the rules of the world, and of melodrama, only to discover that fate, and Hitchcock, have a few uglier tricks up their sleeve. Tricks which Hitchcock has often been too cautious to play. He would have liked to end *Suspicion* with Joan Fontaine dying of the poisoned milk and Cary Grant posting the posthumous letter which will incriminate him. *Shadow of a Doubt* ends with the girl's disillusionment to be imagined— betrays it, indeed, by our relief. *Vertigo* ends, ambiguously, with the spectator able to believe *either* that James Stewart will plunge now into his most terrible depression yet *or* that though his life is emptied of meaning, it is also liberated from a lie. In a sense, Hitchcock is too mediocre to be diabolical. "Stories they wouldn't let me . . . they wouldn't let me . . . they wouldn't let me. . . ." Like a medieval gargoyle, his devil's face is built into, locked impotently in, the outside of the cathedral.

Hitchcock's English world overlaps with that of Greene, of Orwell, with his vision of the semidetached houses as so many "cells of fear." Hitchcock's films constantly tend towards a disillusionment which, far from being liberating, is paralyzing. Hitch as baddie—the sardonic mastermind suggested, yet camouflaged, by the title of master of suspense—seems matched by a fearfulness which has prevented Hitch from expressing his personal vision other than coyly, or in the fates of subsidiary characters, or in the "religious absurdity" of Hitchcock the poet. Hitchcock described to Truffaut his vision of a film about food and what happens to it (as waste, it's flushed into the sea). The vision hardly needs psychoanalysis. Aldous Huxley, I think, first used of Swift that phrase "the excremental vision," which happily describes a certain kind of pessimism (Sade, Huxley himself, Celine, the swamp-and-bathroom syndrome in *Psycho*). It's probably truer to say that Hitchcock's best films leave a nasty taste—so long as the phrase is used admiringly, as it might be of the novels of Gerald Kersh. Kersh is, perhaps, the man Hitchcock ought to be, to be truly Hitchcock—only Hitchcock, quite reasonably, prefers to use his insights in order to tease, to play, to be an affable bogeyman. Kersh is so savage as to be almost unfilmable; yet between Hitch and Kersh comes the early Clouzot. . . .

In the last analysis, perhaps, discussion of religious influences on

Hitchcock are a devious language for discussions of Hitchcock's temperament. For if sect shapes character, it's also true that character chooses sect. The religious absurdity which Hitchcock practices is very different from the moral banalities which his French admirers preach. Dr. Goebbels loved watching *Foreign Correspondent*, perhaps because the weak old statesman who, under torture, makes such a moving speech in favor of the indomitability of democracy, immediately afterwards gives way and does what his captors want. Dr. Goebbels would have enjoyed that to us, and to Hitchcock, pessimistic realism. Hitchcock's *Psycho* and *Rich and Strange* . . . establish the agnostic mixture of pessimism and stoicism, of pain and fearful prudence, which lie beyond, or beneath, a carefully limited existence.

THE FILMS

Notorious

JAMES AGEE

◇◆◇

Notorious lacks many of the qualities which made the best of
Alfred Hitchcock's movies so good, but it has more than enough
good qualities of its own. Hitchcock has always been as good at
domestic psychology as at thrillers, and many times here he makes
a moment in a party, or a lovers' quarrel, or a mere interior
shrewdly exciting in ways that few people in films seem to know.
His great skill in directing women, which boggled in *Spellbound,*
is functioning beautifully again: I think that Ingrid Bergman's
performance here is the best of hers that I have seen. One would
think that the use of the camera subjectively—that is, as one of
the characters—would for many years have been as basic a movie
device as the close-up, but few people try it and Hitchcock is nearly
the only living man I can think of who knows just when and how
to. He is equally resourceful, and exceptional, in his manufacture
of expressive little air pockets of dead silence. He has a strong
sense of the importance of the real place and the real atmosphere;
the shots of Rio de Janeiro are excellent and one late-afternoon
love-scene is equally remarkable in its special emotion and the
grandeur of excitement it gets away with, and in communicating
the exact place, weather, and time of day. Hitchcock also knows the
movie value of the special tones and looks of people with special
backgrounds, at special jobs. Cary Grant, as an American agent, has
almost precisely the cultivated, clipped puzzled-idealist brutality
of a man whom I know in a roughly equivalent job; and Louis
Calhern, as his boss, reminds me even more forcefully that, for
what little good it gets us, there are considerable depths of specifically native sophistication at work, in and out of Washington.
There is perhaps no telling how much of all this should be credited
to Ben Hecht's screen play; but it seems safe to credit a good deal

From Agee on Film: Volume I *(New York: Grosset & Dunlap, Inc.,
1958), pp. 213–14. Copyright © 1958 by The James Agee Trust. Reprinted by permission of Grosset & Dunlap, Inc., and Peter Owen Ltd.*

of the sharpest movie sense, and of a cool kind of insight and control which suggests a good French novelist, to Hitchcock. The story by the way, shows Miss Bergman, a Nazi's daughter and a quondam tramp, doing State Department fingerwork among postwar German plotters in Brazil. Among her more painful duties is marrying Claude Rains, who is no less good as one of Hitler's unhappier orphans than as Bernard Shaw's coldly genial prototype of all dictators.

Three Films
PAULINE KAEL

◆◆◆

The Thirty-Nine Steps (1935)

In 1935 Alfred Hitchcock explained the point of view behind *The Thirty-Nine Steps:* "I am out to give the public good, healthy, mental shake-ups. Civilization has become so screening and sheltering that we cannot experience sufficient thrills at first hand. Therefore, to prevent our becoming sluggish and jellified, we have to experience them artificially." What fun to make a movie in an era when people still needed a bit of a jolt! Even now, oddly enough, these little jolts are more surprising—and certainly more satisfying—than most of the shocks engineered to stun modern audiences. *The Thirty-Nine Steps* is a suave, amusing spy melodrama, directed with so sure a touch that the suspense is charged with wit; it's one of the three or four best things Hitchcock has ever done. The lead, Robert Donat, was that rarity among English actors: a performer with both personal warmth and professional skill. The heroine is Madeleine Carroll. Hitchcock paired them off by the mischievous use of a gimmick: a man and a woman who detest each other are handcuffed together; as day wears into night, they fall in love. The movie thus contains an extra—implicit, as it were

—element of suspense, and theatre employees reported a sharp rise in the use of washroom facilities. Among Hitchcock's pleasing perversities was the casting of Godfrey Tearle, who looked astonishingly like Franklin Delano Roosevelt, as the chief enemy agent. The film also has one of his rare emotionally felt sequences: the brief, chilling scenes between Peggy Ashcroft and John Laurie as a mismated couple, joined together by real chains. John Buchan's novel was adapted by Charles Bennett, Ian Hay, and Alma Reville.

Spellbound (1945)

The idea is intriguing: a murder mystery set among a group of psychoanalysts, with a solution to be arrived at by clues found in a dream which is analyzed. It was carried out in 1945 by one of the most highly publicized collaborations of all time: Alfred Hitchcock and Salvador Dali. The screenplay was by Ben Hecht and the star was Ingrid Bergman, as an analyst, playing opposite Gregory Peck, her amnesia patient—the murder suspect. With all the obvious ingredients for success, Spellbound is a disaster. It's not that Peck's Dali-designed dream life is so bad for Peck, but it's not much for Dali. It was fitting, of course, that the actress who was once described as a "fine, strong, cow-country maiden" should be cast as a good, solid, competent analyst, dispensing cures and murder solutions with the wholesome simplicity of a mother adding wheat germ to the family diet, but Bergman's famous "sincerity" has rarely been so out of place as in this confection whipped up by jaded chefs. Bergman was the great-lady star of the forties; but though she is always a joy to look at and listen to, the sincere apple-checked Bergman is often dull; I prefer the insincere hussy of Saratoga Trunk and Notorious. With Michael Chekhov, John Emery, Leo G. Carroll. Academy Award for Best Original Score (!) to Miklos Rozsa.

Strangers on a Train (1951)

A pretty good case could be made for Alfred Hitchcock as the master entertainer of the movie medium: from the 1930s to the 1960s his films have been a source of perverse pleasure. My favorite among his American films is this bizarre, malicious comedy of 1951, in which the late Robert Walker brought sportive originality to the role of the chilling wit, dear degenerate Bruno. The murder plot is so universally practical that any man may adapt it to his needs: Bruno perceives that though he cannot murder his father with impunity, someone else could; when he meets the unhappily married tennis player Guy (Farley Granger), he murders Guy's wife for him and expects Guy to return the favor. Technically, the climax of the film is the celebrated runaway merry-go-round, but the high point of excitement and amusement is Bruno trying to recover his cigarette lighter while Guy plays a fantastically nerve-wracking tennis match. Even this high point isn't what we remember best from *Strangers on a Train:* it's Robert Walker. It isn't often that we think about a performance in a Hitchcock movie; usually what we recall are bits of "business"—the stump finger in *The Thirty-Nine Steps,* the windmill turning the wrong way in *Foreign Correspondent,* etc. But Walker's performance is what gives this movie much of its character and its peculiar charm.

It is typical of Hollywood's own brand of perversity that Raymond Chandler was never hired to adapt any of his own novels for the screen; he was, however, employed on *Double Indemnity* and *Strangers on a Train* (which is based on a novel by Patricia Highsmith). Chandler (or someone?) provided Hitchcock with some of the best dialogue that ever graced a thriller. With Marion Lorne as Bruno's doting, dotty mother, and Leo G. Carroll, Ruth Roman, Patricia Hitchcock, Laura Elliott.

Notebooks on *Strangers on a Train*
RAYMOND CHANDLER

◆◇◆

Sept. 4, 1950
To: Hamish Hamilton
     . . . I got myself involved in a film job doing a script for Alfred Hitchcock[1] and I don't seem able to do anything else while I'm at it. It's a silly enough story and quite a chore. Why am I doing

From Raymond Chandler Speaking, *edited by Dorothy Gardiner and Kathrine Sorley Walker (Boston: Houghton Mifflin Company, 1962), pp. 132–35. Reprinted by permission of Houghton Mifflin Company and Helga Greene. Title supplied.*

[1] *Strangers on a Train,* from Patricia Highsmith's novel of the same title.

it? Partly because I thought I might like Hitch, which I do, and partly because one gets tired of saying no, and someday I might want to say yes and not get asked. But it won't last beyond the end of this month, I think and hope.

. . . The thing that amuses me about Hitchcock is the way he directs a film in his head before he knows what the story is. You find yourself trying to rationalize the shots he wants to make rather than the story. Every time you get set he jabs you off balance by wanting to do a love scene on top of the Jefferson Memorial or something like that. He has a strong feeling for stage business and mood and background, not so much for the guts of the business. I guess that's why some of his pictures lose their grip on logic and turn into wild chases. Well, it's not the worst way to make a picture. His idea of characters is rather primitive. Nice Young Man, Society Girl, Frightened Woman, Sneaky Old Beldame, Spy, Comic Relief, and so on. But he is as nice as can be to argue with. . . .

EXTRACT FROM NOTES DATED 1950 ABOUT THE SCREENPLAY *Strangers on a Train*

I nearly went crazy myself trying to block out this scene. I hate to say how many times I did it. It's darn near impossible to write, because consider what you have to put over:

(1) A perfectly decent young man agrees to murder a man he doesn't know, has never seen, in order to keep a maniac from giving himself away and from tormenting the nice young man.
(2) From a character point of view, the audience will not believe the nice young man is going to kill anybody, or has any idea of killing anybody.
(3) Nevertheless, the nice young man has to convince Bruno and a reasonable percentage of the audience that what he is about to do is logical and inevitable. This conviction may not outlast the scene, but it has to be there, or else what the hell are the boys talking about.
(4) While convincing Bruno of all this, he has yet to fail to convince him utterly so that some suspicion remains in Bruno's mind that Guy intends some kind of trick, rather than to go through with it in a literal sense.
(5) All through this scene (supposing it can be written this way) we are flirting with the ludicrous. If it is not written and played exactly right, it will be absurd. The reason for this is that the

situation actually is ludicrous in its essence, and this can only be overcome by developing a sort of superficial menace, which really has nothing to do with the business in hand.

(6) Or am I still crazy?

The question I should really like to have answered, although I don't expect an answer to it in this lifetime, is why in the course of nailing the frame of a film together so much energy and thought are invariably expended, and have to be expended, in exactly this sort of contest between a superficial reasonableness and a fundamental idiocy. Why do film stories always have to have this element of the grotesque? Whose fault is it? Is it anybody's fault? Or is it something inseparable from the making of motion pictures? Is it the price you pay for trying to make a dream look as if it really happened? I think possibly it is. When you read a story, you accept its implausibilities and extravagances, because they are no more fantastic than the conventions of the medium itself. But when you look at real people, moving against a real background, and hear them speaking real words, your imagination is anaesthetized. You accept what you see and hear, but you do not complement it from the resources of your own imagination. The motion picture is like a picture of a lady in a half-piece bathing suit. If she wore a few more clothes, you might be intrigued. If she wore no clothes at all, you might be shocked. But the way it is, you are occupied with noticing that her knees are too bony and that her toenails are too large. The modern film tries too hard to be real. Its techniques of illusion are so perfect that it requires no contribution from the audience but a mouthful of popcorn.

Well, what has all this got to do with Guy and Bruno? What a silly question! You shouldn't have asked it. The more real you make Guy and Bruno, the more unreal you make their relationship, the more it stands in need of rationalization and justification. You would like to ignore this and pass on, but you can't. You have to face it, because you have deliberately brought the audience to the point of realizing that what this story is about is the horror of an absurdity become real—an absurdity (please notice because this is very important) which falls just short of being impossible. If you wrote a story about a man who woke up in the morning with three arms, your story would be about what happened to him as a result of this extra arm. You would not have to justify his having it. That would be the premise. But the premise of this story is not that a nice young man might in certain circumstances murder a total stranger just to appease a lunatic. That is the end result. The premise is that if you shake hands with a maniac, you may have sold your soul to the devil.

Sept. 27, 1950
To: Ray Stark[2]

. . . I haven't even spoken on the telephone to Hitchcock since the 21st August when I began to write the screenplay,[3] which was written in one day over five weeks. Not bad for a rather plodding sort of worker like myself. I don't know whether he likes it, or whether he thinks it stinks. The only method I have of deducing an answer to this question is that I was allowed to finish it. . . . Some of the scenes are far too wordy, partly because, as Woodrow Wilson once said, 'I didn't have time to write it shorter,' and partly because I didn't know what Hitch was doing to the script himself. . . . It must be rather unusual in Hollywood for a writer to do an entire screenplay without a single discussion with the producer. . . .

[2] Chandler's representative at the time.
[3] Strangers on a Train.

Strangers on a Train: The Pattern of Encounter
RONALD CHRIST

◈◈

To talk or read about Strangers on a Train is often to encounter laborious and finally wearisome interpretations that take little account of the aesthetic incarnation that is the film when we view it: the notion of moral complicity, the choreography of a suite for doubles have more to do, it seems, with a "meaning" than they do with sensuous artifice. And so Hitchcock's invention is reduced to the quality of perception exhibited in Patricia Highsmith's novel or countless other variations on the theme of the psychological double or of pervasive evil. Of course not all critics, not Truffaut or Robin Wood, for example, entirely neglect the complex thing that the movie shows for the models or ideas it embodies (although none seems concerned with the feelings, other than suspense, it arouses).

But if any do, the blame or at least the responsibility must rest squarely with Hitchcock. He has designed an image so symmetrically patterned—"mapped out like a diagram," says Truffaut.[1] Explicitly so. But oddly enough Hitchcock's surest gesture in the direction of the interpreters (and thus perhaps his most blameworthy) has been ignored by them, indicating once more that in being intent on rendering account, we fail to see.

I am referring to that scene near the suspenseful climax of the film where Guy and Ann are walking down the steps of the stadium at Forest Hills. Guy, you remember, must beat his opponent rapidly in order to catch the train for Metcalf in time to prevent Bruno from planting Guy's lighter at the spot where Miriam was killed. As Guy and Ann descend the stairs, we can see, near the top of the screen, these partially obscured lines written in Gothic letters across an ornamental placard:

If you can . . .
And treat those two impostors just the same.

The thematic reverberations of this conditional injunction are immense for anyone working out the double/moral implications of the imagery of doubles: The inscription refers by visual association to Guy and Ann, *not* to Guy and Bruno, although the word "those" (instead of "these") perhaps points beyond the scope of what is being shown on the screen at that time.[2] But more importantly, the inscription serves as a guide, not so much for "treating" or interpreting of any two characters, as it does for taking an attitude toward the whole film. The inscription is a glaring intrusion; it is more flagrant in its import than Hitchcock's customary appearance in his films. Those comic signatures are worked into the films as painters may embody the message of their initials in the medium of their design. Moreover, those appearances have lost meaning through mandatory reoccurrence—they are something to be gotten over with as soon as possible, as Hitchcock himself admits —whereas the writing on the wooden placard is commanded from outside the world in which the characters live with an eschatological force that Hitchcock's significantly silent appearances never have. If you did not know of Hitchcock's custom and could not recognize his face, your seeing him in his own films would not affect your attitude toward those films; but that inscription in *Strangers on a Train* is entirely self-sufficient and therefore vastly more authori-

---

[1] François Truffaut, *Hitchcock* (New York: Simon & Schuster, 1967), p. 146.
[2] [See Hitchcock's remarks on doubles in the interviews with Bogdanovich, this volume, pp. 30–31. See also André Bazin in this volume, pp. 66–67.]

tative. It jolts us from illusion and forces us to see that the moral
issue is not a matter for our discovery and that the film is not the
process of its disclosure. In other words, what Hitchcock accomplishes with that inscription is the succinct communication of a
theme and an attitude, thereby precluding the necessity for interpretation. He accomplishes this by means of an ironic, cinematic
parabasis:[3] The Gothic characters spell out Hitchcock's tone and
the device frees us to see the *body* of this work, since its *spirit* is
already revealed.

Freed from the obligation to moralize and psychologize, perhaps
even literally put off by the blatant and humorous disregard for
authenticity[4] evidenced in this directorial graffito, we are encour-

---

[3] *Parabasis:* in Greek Old Comedy, that section of the play where the chorus
turns to the audience and addresses them directly in the interest of the author,
explaining his point of view, his technique, and even his difficulties in real
life. In its "nondramatic" function and in its illusion-breaking force, the *parabasis* foreshadows much that seems peculiarly modern in drama. Hitchcock's signature appearances sometimes resemble the convention, as when his picture
shows up in a before-and-after advertisement for some reducing product on the
back pages of a newspaper being read by one of his characters.

The source of the quotation is the second stanza of Kipling's poem "A Charm"
in his volume *Rewards and Fairies* (1910) and reads in its entirety:

> If you can meet with Triumph and Disaster
> And treat those two impostors just the same.

Given that source and that good-sportsman tone, it is just possible that the
quotation is a motto of Forest Hills or of some other stadium which actually
provided the location for the scene. Critics I have discussed the point with have
thought so; sports *are* stranger than movies it seems. But it is my sense that the
scene in the stands was not filmed on location. Certainly it is the force of the
filmed scene—by obscuring the abstractions "Triumph" and "Disaster"—to
point away from the quotation's applicability to a sporting event and to focus
the words exclusively on the filmed drama. If the words do belong to the "real"
context of the stadium, therefore, the movie undercuts that relationship—
obliterates it, I would say—in a way that is not true of Hitchcock's subsequent
use of contextual mottos.

[4] The matter of *authenticity,* and especially scenic authenticity, in Hitchcock's work requires further observation. His use of trick and process shots
often implies, as in *Torn Curtain,* an ironic commentary on the foreground
action; often establishes, as in *Shadow of a Doubt,* a corrupting contrast to the
location shots; and almost always creates, as in *Strangers on a Train* (where the
dog licks Guy's hand in slow motion), an awkward, staged effect superficially
incompatible with the clever technical devices—the stairwell scene in *Vertigo,*
the illuminated glass of milk in *Suspicion,* the apocalyptic carousel in *Strangers
on a Train*—and certainly produces an effect that calls our attention to the
creator behind the scene. These artifices, which often seem only artificial, are
clearly an element of his style, no matter what production motives he may
adduce as their cause. Hitchcock discusses "authenticity" with Truffaut (*Hitchcock,* pp. 177–79), but to little purpose since what he understands by the word
in that interview is something like "accuracy in reconstructing a historical
event."

aged to take a more detached view of the emotions expressed by the characters—a concern of the actors—and witness the pattern of phenomena—the affair of the director. We are able to see that for Hitchcock signification is easily attained and therefore properly thrown away, tossed off in a secretive gibe, whereas manifestation is the joy of the director to invent and of the viewer to discover. Thus a more proper analysis of *Strangers on a Train* might assume the moral premise and the psychological theory encapsulated in that inscription and devote itself to the scheme whereby Hitchcock materializes those abstractions.

The crucial locus of patterned phenomena in *Strangers on a Train* is the intersection of dialogue and visual scene. For example, when Guy and Bruno are talking, early in their acquaintance on the train, Bruno orders some drinks—a couple of Scotches—but what he actually says is "a pair of doubles"—exactly what we are looking at. Nowhere else in the film, or in any of the criticism of it for that matter, is Hitchcock's theme so swiftly indicated exclusively by means of what is simultaneously seen and heard on the screen. However, for the special twist Hitchcock gives to that theme, a subsequent scene is even more significant: Just after Guy has left Bruno's compartment on the train and Bruno is still holding Guy's cigarette lighter, Bruno repeats to himself: "Crisscross." Here is the gist of the plot: the exchange of murders Guy and Bruno have discussed, the ambiguous exchange of guilt and identity (for example, Ann's sister becomes Miriam for Bruno, Guy becomes the murderer to the police). Here is the emphasis of the movie in contrast to the book, where the word "crisscross" occurs only incidentally in the description of a fence.[5] Here too is the foundation for the particular inflection Hitchcock imposes on his theme. That inflection is made verbally evident when Guy responds to Miriam's telling him that she will not get a divorce. He exclaims: "You little doublecrosser!" But then the inflection is equally evident in Guy's trying to inform Mr. Anthony of Bruno's plot and in Bruno's attempt to plant the lighter on the island—both doublecrosses.

The metaphorical signification of the two words "crisscross" and "doublecross"—their narrative and thematic bearing—is clear. But the way their geometric denotation informs the figure Hitchcock has invented needs further observation, because the complementary phrases gather a number of apparently disparate scenes—such as Bruno and Guy's entrance into the railroad station, the loss of Guy's lighter on the train, the tennis matches, and Bruno's return

---

[5] "He walked through the wide gate between the crisscross wire fence and looked up College Avenue again." Patricia Highsmith, *Strangers on a Train* (New York, 1950), p. 32.

to Metcalf—into a rhythmic pattern of criss- and doublecrosses which organizes the perceptual experience of the film.

In the famous opening of the film, we see a pair of feet going in one direction and then a pair of feet going in the other. Specifically, Bruno, moving through space diagonally, crosses from the viewer's right to the viewer's left, then Guy moves in the opposite direction. The sequence is repeated twice and then the two men, one after the other, walk straight back, away from the audience, into the inner part of the station. Although no literal crisscrossing is seen, we translate the sequential pattern of feet describing one diagonal arm of the cross in space at one time and the other arm at a subsequent time, into a unified, simultaneous pattern. This displaced pattern is confirmed when we see the crisscrossing railroad tracks. The pattern in the actions of Guy and Bruno is therefore apparent to us before they are aware of it and long before it is actually apparent on the screen: The rails are the explicit confirmation of the implicit diagram of their walking.[6] And if we look at those rails closely, we see that Hitchcock again shows not one intersection —a crisscross—but two—a doublecross. This intensified pattern is corroborated in the characters' feet, which collide when Guy's *crossed* leg knocks Bruno's *crossed* leg—a "double crossing." The visual doubling is reinforced by having Guy and Bruno hold their dinner conversation in front of a train window, which faintly and eerily juxtaposes their reflections to their screen images (a visual precursor of the murder scene, which is shown in the distorted, reflected image on Miriam's glasses) at the same time that their shadows—a traditional manifestation of the double—are projected on the walls adjacent to the window. Their talking heads are located explicitly between their dumb, blurred, dark shadows and their silent, gesturing, ghostlike reflections.[7]

Once Bruno has examined Guy's lighter, Hitchcock makes clear

---

[6] The relationship between the walking feet and the veering rails is rightly taken for granted by Truffaut: "One of the best things in *Strangers on a Train* is the exposition, with the follow shots on feet going one way and then the other. There are also the crisscrossing rails" (*Hitchcock,* p. 144).

[7] A still of this scene, which is only on the screen briefly, can be studied in Truffaut's *Hitchcock* (p. 143), but the shadow motif should be observed throughout the film; most notably, perhaps, in the Tunnel of Love scene where Bruno's shadow overtakes and passes those of Miriam and her boyfriends; again—more subtly, but no less melodramatically—in the scene where Bruno rises up in bed to confront the sneaking Guy. Hitchcock also manages to draw other characters into the pattern of the piece by use of the shadow motif. For instance, Ann's sister, says to Guy (about one of the pair of detectives who follow him): "I've just been talking to your shadow."

For comment on the connection of shadows and doubles see Otto Rank, *The Double: A Psychoanalytic Study* (Chapel Hill: University of North Carolina Press, 1971).

that the crisscross pattern is not only the condition of their encounter but also the shape of Guy's destiny, for the lighter has a
pair of embossed, crisscrossed tennis rackets in the upper left-hand
corner, and the lighter is a gift from Ann (the initials "A" and "G"
appear on the lower left-hand corner). Of course, when he gives
the lighter to Bruno, Guy reinforces the pattern by relinquishing
the object of his identity to the man who has proposed the recipro-
cal murder. Here, in words and images, is a real exchange in the
film: the lighter for the proposal, the object for the words. And the
heft of Bruno's character, delicate as it may be, is suggested by the
fact that at the end of the scene he retains both tokens of identity:
Guy's lighter and his own tie clasp, which spells out his name in
round, silver letters and is a gift from his mother.

The crossed tennis rackets are the emblem of the game and of
Guy himself. Tennis is therefore established as a crossing of op-
ponents, equivalent, in a miniature way, to the crossing of Guy
and Bruno. Therefore, Hitchcock subtly rhymes the tennis matches
with the opening footage. In the scene at Guy's practice session,
where Bruno looks straight at Guy as the spectators' heads move to
right and left in diagonal lines of vision, and in the final match,
where the play is necessarily sequential, we are watching implicit
crisscrosses like those we saw in the exposition. The arms of the
cross are described not simultaneously, but rather sequentially as,
for example, when the ball is hit from right backcourt to left
forecourt or when the cutting of the scene forces us to see the
players in isolation from one another, even though they are locked
together in a spatial pattern—just as the pairs of feet were at the
beginning of the film.[8]

But since the lighter with the crossed rackets is passed on to
Bruno, it becomes his emblem too, and the cross appears in a spatially integrated form when Bruno arrives in Metcalf for the second
time. Just before he drops the lighter down the drain, we see him
standing on the curb, and in the distance—at the level of his head
on the screen—we see a railroad sign. The composition is like that
of a medieval painting showing a martyr and his device. The sign
is a familiar one to us all—a "stop-look-listen"—and we do not fail
to take it for a sign because it is, of course, in the shape of a crisscross, symbolizing a train crossing. The words on the sign, unlike
those on the stadium placard, come from *within* the world of the
characters. These words are no less ironic in their direct applicability to Bruno at that moment, and their format is no less consistent with the paradigm Bruno himself enunciated than are Guy's
lighter or the tennis matches, but the initial association of that

---

[8] [See also Andrew Sarris's comments, this volume, pp. 89–90.]

railroad crossing with the crisscross pattern has been emphatically established by Hitchcock's showing us the Metcalf crossing immediately after Bruno's utterance. From the word "crisscross" we moved without interruption to the vision of the train crossing. Nevertheless, Bruno's last gestures are in fact aimed at rejecting the emblem by relinquishing the lighter and reattributing it to Guy (and thus attributing the murder to him too). In this light his dropping the lighter down the drain is expressive of his intention, but his retrieving it is evidence of his fate. Ultimately, the pattern is dominant and Bruno dies with the lighter clutched tightly in his hand; and then, once dead, he lies revealed, stigmatized with it in the palm of his hand.

In order to complete his pattern, however, Hitchcock does not end the film with Bruno's death. Instead he shows a final scene with Ann and Guy on a train in the presence of a minister, who executes an ironic, humorous doublecross to Bruno's crisscross by unwittingly doubling Bruno's lines from the earlier train conversation: "Aren't you Guy Haines?" This last scene, like so many other scenes in the film, is the matched double of an earlier one in which Guy, on his way to Metcalf to intercept Bruno, witnesses two men strike up a conversation when their crossed feet accidentally touch. In the scene with the two men, Guy interrupts his questioner by moving away and the visual-auditory pattern—already weakened because one of the two men in the earlier scene did not, could not ask, "Aren't you Guy Haines?"—is broken: The film ends.

The tone of that final scene is humorous, and its effect, unlike the effect achieved by other recurring scenes—the ending of *All about Eve,* for example—is to make us disbelieve in the possibility of the situation being repeated. The conclusion, then, is not moralistic or even fatalistic, but rather an expression of Hitchcock's attitude toward his subject: The tonality of the ending is like the tonality of the intrusive writing on the stadium placard or like that of Hitchcock's signature appearance, in which the essential theme and pattern of the film are parodied by the director's having Guy and Hitchcock encounter each other in a crisscross pattern on the train steps as Hitchcock climbs aboard carrying a double bass, the visual repetition of his own form. These scenes return us to the shape of the film—the fact and shape of recurrent intersection—more than they do to the meaning of recurrence, because, in fact, there is no meaning in the minister's repeating Bruno's words or in Guy's inconsequential passing by Hitchcock except the literal signification. Hitchcock guarantees our laughing (snickering, really) at the *form* of recurrence divested of material motive precisely by making his meeting with Guy lead to nothing, and by having the man a minister of urbane and ineffectual costume and countenance. In

other words, the conclusion of the film belongs more to the completion of the pattern—the sense of completion itself—than it does to
any congruent meaning. The conclusion is comic and suggests that
play with forms which is characteristic of comedy, that play I have
been trying, with apparent and necessary failures, to isolate in its
relative purity in *Strangers on a Train.*

The Wrong Man
ERIC ROHMER and CLAUDE CHABROL [1]

❖❖

While waiting for the future joys which Alfred Hitchcock won't
fail to give us, it is pleasant to . . . study . . . a film which not
only binds up themes scattered in all the works, but which elo-
quently proves that it isn't a waste of time to want to deeply illumi-
nate it as well.

The Paramount series and the television programs have made
Hitchcock more popular than he ever has been. Feeling strong
enough to face eventual commercial rejection, he decided to make
a film he believed in. It is for Warner Brothers that he produced, in
the beginning of 1956, *The Wrong Man,* settling for 10 percent
of the box-office receipts (which he knew would be problematic).

This work of obvious ambition is devoid not only of imaginative
fiction but of all "suspense" too.[2] It is, like *Lifeboat,* a fable, but at
the same time the exact narration of a piece of news. Is this a coincidence? The genre of allegory, most often a pretext for mediocri-

From Hitchcock *by Eric Rohmer and Claude Chabrol (Paris: Editions Universitaires, 1957), pp. 151–59. Copyright © 1957 by Eric
Rohmer and Claude Chabrol. Translated by Stephen Arkin, Alan
Schiffmann, and Albert J. LaValley. Used by permission of the publisher. Title supplied.*

[1] [For a discussion of the Rohmer–Chabrol approach, see the selections by
Robin Wood, this volume, pp. 76–78, and Raymond Durgnat, pp. 91–96.]

[2] Hitchcock seems to have temporarily given up comedy at the same time as
he has given up John Michael Hayes. Maxwell Anderson wrote the dialogue for
The Wrong Man and he will write *D'Entre les Morts (Vertigo),* based on the
novel of Boileau–Narcejac. [The screenplay was not written by Anderson but
by Alec Coppel and Samuel Taylor.]

ties, is the same to which the most original recent films belong: *Le vent souffle où il veut, Voyage to Italy (Strangers), Mr. Arkadin, Eléna et les hommes.* And Bresson, Rossellini, Welles, and Renoir were able, with the same success as Hitchcock, to handle the apparently contrary capacities of the allegorical and the documentary form (even if it were only, as in the case of *Eléna,* the documentary of an era). Concrete reality furnishes the story with the flesh, without which it is nothing more than an exercise.

It is with the documentary that we will begin. If *The Wrong Man* were nothing more than an accurate record of the "police scene," the title alone would suffice to legitimize our admiration. One man is arrested instead of another, and because he is innocent, enjoys at a distance what he suffers, even though he suffers all the more because of that distance. Hitchcock, who likes to drag his characters in the mud of contempt, finds the most adequate form for the expression of this contempt. His fascination with the abject, to which he yields and makes us yield at the same times as his hero, can't help but recall Murnau's *The Last Laugh.* Christopher Emmanuel Balestrero (Henry Fonda), bass player at the Stork Club, comes to the office of an insurance company to borrow on his wife's policy. He is identified as the perpetrator of a number of holdups that have taken place in the previous few months, and the police pick him up just as he is about to open the door of his house. During the entire first half of the film, we do not stop following him through the slow formalities—one might call it the ceremony—of the police and the authorities. There is, first of all, the confrontation with the witnesses. Now under the obstinate stares of the office employees, now under the barely curious stares of the shopkeepers in front of whom he is made to march, he is nothing more than the stranger one stares at, the thing one identifies. Then there is the comparison of handwriting, during which we are spared neither the time the accused takes to write nor the monotonous scratching of the pen on the paper. There is the booking (many times in the course of the film one will hear him recite his name, his age, his profession: "Christopher Emmanuel Balestrero, 38, musician."). There is the shame of being put in a cell, the greater indignity of being handcuffed, the prompt routine of the accusation and the enormity of the bail, the entrance into prison, the promiscuity of the shower, the lockup.

A documentary without flourishes, though it is never shown to us by an impassive camera, but a descriptive, narrative one, as Hitchcock's always is. And the truthfulness of detail only serves to better support the force of the symbols. We not only clearly discover the idea but, if we may say so, we feel it. This idea furthermore is very complex. It is an idea that we can successively identify

as: the fundamental abjectness of a human being who, deprived of
his freedom, is nothing more than an object among other objects;
misfortune at the same time unjust and deserved, like that of Job
(everything seems to conspire against our musician); guilt just as
fundamental as that which serves as a theme in Kafka's *The Trial*.
And, when Balestrero, who understands the futility of all protest,
sees himself reduced to the role of spectator at his own downfall,
another motif, that of redemption, emerges with the others. The
face and Christlike position of Fonda in his cell remind us of the
iconography of the Stations of the Cross.

The prisoner is freed, his brother-in-law having collected the
necessary amount for bail. From this moment on, it is allegory which
dominates. Everything, we have said, seems to conspire against him.
To an almost unbelievable point. His lawyer having told him to
find witnesses, he successively learns of the death of the only two
who can furnish him with an alibi. His wife then goes mad. Of all
Hitchcock's films, this is certainly the least fictional at the same time
that it is the least believable. This is why the director has taken
great pains to establish for us the authenticity of the story. A
modern application of a principle dear to Corneille: knowing that
an event out of the ordinary can produce tragedy, on the condition
that it is "possible." And the proof, *a fortiori*, that it is possible is
that it has happened.

The extraordinary here is not then, as in the preceding works, a
simple procedure, a pretext for brilliant developments. It appears
in and for itself and becomes the very object of the study. Thus it
was crucial for Hitchcock to grasp it in this privileged, accomplished
form, which is exactly that of the miracle. (*Le vent souffle où il veut,
Voyage to Italy* [*Strangers*], and *Mr. Arkadin* also showed us miracles.) Balestrero is on the verge of being condemned, but the im-
promptu interruption by a member of the jury, tired of the lawyer's
finickiness, causes a mistrial. Once again, he finds himself free but
alone (his wife has been placed in an institution). More desperate
than ever, on the advice of his mother he begins to pray. Then, while
he is contemplating the image of the Sacred Heart, a split image
shows the guilty man walking down the street advancing towards
the camera until his face meets that of Fonda. The man will get
caught himself a few moments later, and Balestrero will be cleared.

Has there been a real miracle? Nothing enables us to deny it,
but, as opposed to what happens in Dreyer's *Ordet* and despite
the obvious prejudices of the narrator, a certain freedom of judgment has been left us. Nevertheless, there is no desire on Hitchcock's part to deride this idea of providence encountered in our
path. What the director denounces is, on the contrary, the weak
surrender to chance (it is significant that our wrong man plays

the horses in his spare time). What he condemns even more severely are the two theological sins of pride and despair. The misfortune which afflicts the hero and his wife is extraordinary only because they are both willing to believe that it is extraordinary. If we weigh objectively what happened to them, we find in the balance as much good (the payment of the bail and the dismissal of the trial) as bad. They are both victims of their own lack of trust in divine benevolence and the virtue of their own will. They give in to the fascination of diabolic scheming (that other Hitchcock theme). Finally, the last mistake of our false victim, who is really a false innocent like all men since the Fall, is to believe that because of one miracle a second is due them. He seems to imitate that legendary shepherdess who threw herself down a precipice in order to escape the local ruler and was miraculously saved. When she tried to repeat the exploit, she killed herself. But his wife will remain mad —at least for the time being—and as the nurse suggests, human patience is necessary for her recovery, even though we are given reason to believe that there too a miracle would be involved.

The conclusion is ambiguous, but it is not a dodge. This ambiguity is in things themselves. It is characteristic of Hitchcock to show us simultaneously the inside and outside of things. His work moves between two poles, which, like extremes, can coincide. We have used the term "exchange" for this movement and we must admit that it finds here its most noble expression in the idea of the exchangeable guilt of mankind. In retrospect this idea adds a new wealth and depth to what had appeared modest and superficial elsewhere.

As for the form, its basis is perhaps more difficult to locate, but it is no less rigorous than in the works of supposedly pure virtuosity. The dominant image will be, naturally, that of the wall. Hitchcock is always able to take advantage of the very accident of filming and profits from the fact that he cannot, in a real setting (which was most of the time that of the original news item) allow his camera to back away very far. He stays close to the faces and thereby increases our sense of suffocation. The fundamental chord is struck (after a brief prelude at the doors opened and then closed) by the scene in the office of the insurance company. On the other side of the counter the employees eye Balestrero as they hide behind the figures of their neighbors. Also, when the musician is taken by the police in a taxi, his view is blocked by the faces of the detectives. The first two faces in profile, and the third of whom he sees only the forehead and eyes in the rearview mirror. This last framing, this frame in the frame, reappears in the form of the window of the prison door. The camera penetrates through it and later, when it is announced that the bail has been paid, the camera leaves, isolating the two enormous eyes of the prisoner. Artifice? The noble art of a filmmaker who knows, as a worthy emulator of Murnau, how to brutally dehumanize the face which he otherwise shows to be human, too human, and thus bracket for an instant the presence of the soul in order to make it more evident elsewhere, and finally to humiliate a man in that which is most noble, his gaze. This interpretation is confirmed by the broken mirror in which the image of Fonda is reflected and deformed as the director says, "like a Picasso." In the hands of a director of genius (and we can now use this word without fear) the usual classic form of the film can perform feats which the members of the so-called "avant garde" have failed to perform and will fail to perform. And it is Stravinski whom Hitchcock evokes again *a propos* the same scene, concise beyond belief, in which Vera Miles hits Fonda. We see only the start of the blow and its conclusion, an abruptness which accentuates the strangeness of her mad gesture and makes all the more oppressive her slow return towards the bed.

The most diverse styles make a most happy mixture in this film and their successive appearance does not break its perfect homogeneity at all. We are spared many intermediaries (spatial or temporal) but certain moments, apparently unimportant, are presented in their precise duration, for example, the already mentioned scene of the handwriting analysis, or when Balestrero's wife calls the lawyer on the phone. The point of view is subjective only in appearance. Although we see with our own eyes (at the time of the handcuffing when the shape of his shoulder appears, or when he dares not look at his fellow prisoners and sees but a row of feet on the floor of the paddy wagon), the hero remains exterior to us as he does to himself. This false subjectivity, this false exteriority are really in the basic tenor of the film. To the ambiguity of content corresponds a constant ambiguity of form. Which proves once again that the procedure itself is nothing: what counts is what is done with it. The scene of the prisoner in his cell owes its tragic grandeur to the most simple means (a few glimpses of the walls and ceiling). For someone else they would have been dull and old-fashioned, just as that extraordinary pan could appear old-fashioned on paper (the one which ends the scene when the camera in a dizzy movement performs a circular dance in front of Fonda).

Once the prisoner is free, the impression of suffocation gives way to one rather similar to the conspiracy of occult forces. Balestrero runs through the city and the countryside in search of evidence, but his territory is always limited. We see him either at the village inn, coming up to an empty table, or he opens a door, from behind which jump out like a jack-in-the-box two astonished and laughing young girls, interrupted in their amusement.

In this film of night, this film of winter, this film of black and white, where Robert Burks adapts the tonality of his photography to that of the subject with no less ease than in the preceding works in color, the gaze is emphasized as it has been in *I Confess*. The stubborn and self-satisfied gaze of the witnesses, the professional gaze of the police, the lawyer, the psychiatrist, the mad gaze of Vera Miles, the gazes of Henry Fonda (to describe them would be to recount the entire film again), and finally, the gazes of confrontation between the true "guilty figures," by which the gazes become "transmission wires" of the exchange. The first passes his guilt to the other. The sound of objects, the sound of voices off camera are done carefully. The noise of the elevated subway introduces a leitmotif, and the beautiful score of Bernard Herrmann blends with the general austerity of the work.

. . . As a final feature let us simply choose among all the aspects of a genius of a thousand faces the one which appears to be the most indisputable. Hitchcock is one of the greatest inventors of form in the entire history of cinema. Only Murnau and Eisenstein can on this count be compared to him. Our task will not have been vain if we have been able to show how, starting with form and working with its rigor, an entire moral universe has been elaborated. Here the form does not beautify the content, it creates it. Hitchcock can be summed up in this formula. This is what we wished to prove.

Hitchcock, Truffaut, and the
Irresponsible Audience
LEO BRAUDY

❖❖

In the beginning of his opulently mounted interview with Hitchcock,[1] François Truffaut writes that Hitchcock has always feared

From Film Quarterly *21, no. 4 (1968): 21–27.* © *1968 by The Regents of the University of California. Reprinted by permission of the author and The Regents.*

[1] François Truffaut, *Hitchcock* (New York: Simon & Schuster, 1967).

technicians who might "jeopardize the integrity of his work." But in this "definitive study" (to cite the dustjacket) Truffaut's own approach is so doggedly technical, so intent on style as opposed to meaning, that one wonders if the feared technicians haven't come in by a rear window after all. The interview is an anatomy of Hitchcock's work that shows little sense of what technical methods signify or what stylistic devices express. Truffaut draws back from any exploration of the psychological depths of either Hitchcock himself or the movies Hitchcock has made. Hitchcock makes many leading remarks about his themes and methods that Truffaut glosses over. Hitchcock reveals fascinating shards of his psychological night-life, but Truffaut only alludes to the dark area of voyeurism, ex-hibitionism, and fetishism that Hitchcock's films explore; he is too interested in showing his own knowledge of plot and technical details to go any further. And because of his lack of interest in the psychological dimensions of Hitchcock's films, Truffaut misses how Hitchcock in his best films manipulates the deepest reactions of his audience.

Has Truffaut been hampered by the difficulties of a long interview (fifty hours spread over several days), complete with translator? If we cannot have the experience of two directors talking equally, let us have an incisive picture of one. But Truffaut gives us neither. Recent journalism has developed the interview into a vehicle of self-revelation. But what we learn about Hitchcock from *Hitchcock* is less due to Truffaut's insight than to his inclusiveness. There are 472 stills and full credits for all of Hitchcock's films. There is even a developing plot relation between two characters named "Hitch-cock" and "Truffaut" which can be followed as a welcome counter-point to the more obvious play of question and answer. But this plot reveals neither Truffaut nor Hitchcock; each tries to direct and each has cast the other in an uncongenial role. Truffaut's early impulse is to score points. He shows that his memory of *The Last Laugh* is better than Hitchcock's and he tries to make Hitchcock admit that his work was influenced by Fritz Lang. Hitchcock re-sponds with his usual mask of evasive humor: he can't remember *M, The Spy,* or *The Testament of Dr. Mabuse,* but he will admit to changing a scene in the first version of *The Man Who Knew Too Much* because he had noticed a similar scene in Mervyn LeRoy's *I Am a Fugitive from a Chain Gang.* Underground arguments some-times flare. While discussing *The Ring,* Hitchcock mentions visual touches he thinks no one noticed; Truffaut nods but wants to talk about what *he* noticed; Hitchcock replies that all the reviewers noticed *those* details. None of these conflicts is more than trivially illuminating. And it is difficult not to find Truffaut at fault. Instead of facing Hitchcock with probing questions, he plays the eager young man, ready to reel off complicated plots the master has forgotten, adulatory and bumptiously arrogant at the same time. Instead of drawing Hitchcock out, Truffaut forces him back into his old masks.

Ideally, an interview can be a process of understanding. But Truffaut has certain set ideas about Hitchcock. His emphasis on Hitchcock's technique of suspense and "dramatic impact" shows traces of the same kind of condescension or reverse snobbery that dubs Hitchcock "the world's foremost technician": however great a director Truffaut believes Hitchcock to be, he may not expect him to be interested in psychological themes as complex as those dealt with in *Jules and Jim*. This bias leads naturally to Truffaut's concern with workmanship and technical detail. He calls *Notorious* "the very quintessence of Hitchcock," "a model of scenario construction." Hitchcock calls the single-shot technique of *Rope* "quite nonsensical," but Truffaut's questions follow the familiar litany: "What about the problems with the color?" "What about the problems of a mobile camera?" "What is truly remarkable is that all of this was done so silently that you were able to make a direct sound track." Faced with Truffaut's almost programmatic bias, Hitchcock finds he can respond only in Truffaut's terms, and in the latter part of the interview he finally asserts—with Truffaut's approval—that he likes technical tricks much more than subject matter or acting.

Hitchcock's seeming agreement with Truffaut rests actually on a very different definition of technique that uses, however, much of the same language. Both Truffaut and Hitchcock make oddly archaic statements about the way sound film ended the great era of the cinema. Truffaut seems to have forgotten André Bazin's attacks against "pure cinema" cultists (such as "The Virtues and Limitations of Montage") for he comes on like young Raymond Spottiswoode. In line with his interest in technical details and fragments of directorial style, he treats each film as a "pure" object: a compound of techniques, or problems solved and unsolved. But all of Hitchcock's "techniques" are aimed at destroying the separation between the film and its audience. When Truffaut talks about the emotional effect of a film, he is speaking of dramatic irony, surprise, and the shock of realism. When Hitchcock talks about emotion, he is asserting the audience's involvement and implication in what is happening on the screen. In speaking of *Psycho,* Hitchcock appears to follow the "pure" cinema line: "It wasn't a message that stirred the audience, nor was it a great performance or their enjoyment of the novel. They were aroused by pure film." Truffaut answers, satisfied, "Yes, that's true." But Hitchcock explains further what he means: ". . . The construction of the story and the way in which it was told caused audiences all over the world to react and become emotional." Truffaut responds: "Yes, emotional and even physical." Hitchcock snaps: "Emotional."

In the first half of the interview Hitchcock frequently drops hints of some larger issues, but Truffaut, bound in his own interests, plows on. Hitchcock suggests, for example, that his use of handcuffs has "deeper implications":

> *A.H.* Being tied to something . . . it's somewhere in the area of fetishism, isn't it?
>
> *F.T.* I don't know, but I have noticed that handcuffs have a way of recurring in your movies.

While Hitchcock vainly implies the emotional and psychological relevance of his details, Truffaut concentrates on an intellectualized appreciation of fine finish and professional gloss. He says of the death of Mr. Memory in *The Thirty-Nine Steps*: "It's this kind of touch that gives so many of your pictures a quality that's extremely satisfying to the mind: A characterization is developed to the limit —until death itself." Truffaut therefore interprets the paranoia implied by the subjective camera in *The Thirty-Nine Steps* in technical terms as Hitchcock's effort "to sacrifice plausibility in favor of pure emotion." He does not perceive the relation between Hitchcock's typical technical devices and his deepest thematic concerns.

Truffaut's analysis and questioning falls down therefore whenever he touches upon larger areas of structure and meaning in Hitchcock's films. Truffaut dispenses with plot in the name of "pure" cinema; Hitchcock cares little about the minor springs of plot— what he calls the "MacGuffin," the gimmick—because he is dealing with more inclusive rhythms. "To me, the narrator, they're of no importance." And this narrative sense, Hitchcock asserts, despite Truffaut's concern with technical virtuosity, is the most important part of his directional method. Truffaut talks about technique, but Hitchcock talks about the audience and its psychology. He manipulates the audience for his own ends, and he wants them to leave his films with a narrative sense of what has occurred. Truffaut does not grasp this idea because each film is for him a pure aesthetic object. But for Hitchcock it is the medium for a relation between the director and the audience. Truffaut discusses camera movement in terms of "dramatic impact," but Hitchcock continually expresses it as an element in establishing point of view.

Because of Truffaut's inability or unwillingness to explore Hitchcock's interest in point of view and his skirting of psychological themes and preoccupations, he is particularly blind to the central area of Hitchcock's work where technique and theme coincide in

the study of voyeurism. Building on the interplay between directorial construction and audience understanding that is the basis of
montage, Hitchcock develops certain themes that rely directly on
the experience of watching a film itelf. Even when Truffaut touches
on the theme of voyeurism, he believes that the psychological interest is fortuitous:

> *F.T.* Would you say that [James] Stewart [in *Rear Window*] was
> merely curious?
>
> *A.H.* He's a real Peeping Tom. . . . Sure, hes a snooper, but
> aren't we all?
>
> *F.T.* We're all voyeurs to some extent, if only when we see an
> intimate film. And James Stewart is exactly in the position of a
> spectator looking at a movie.
>
> *A.H.* I'll bet you that nine out of ten people, if they see a woman
> across the courtyard undressing for bed, or even a man puttering
> around in his room, will stay and look; no one turns away and says,
> "It's none of my business." They could pull down their blinds, but
> they never do; they stand there and look out.
>
> *F.T.* My guess is that at the outset your interest in the picture
> was purely technical, but in working on the script, you began to
> attach more importance to the story itself. Intentionally or not, that
> back yard conveys an image of the world.

All through the interview Hitchcock has made remarks about "Peeping Tom audiences" and his efforts to manipulate them. But Truffaut never sees the larger thematic and structural implications of
this interest.

Every movie is naturally voyeuristic, not only the most intimate
ones, and that is a great part of their appeal—the sensuous imme-
diacy that goes beyond the stylized realism of the fourth-wall theater.
A feeling of occasion and artifice may separate us from a particular
movie, as it usually separates us from even the most realistic play.
But with the camera eye substituted for our own the potentiality
for greater intimacy, mediated by "me, the narrator," is still there.
The films of Hitchcock play in different ways with these psychological assumptions of the film form itself. Some are less successful and
perhaps deserve the technically oriented analysis of Truffaut. But
voyeurism is more than a metaphor for Hitchcock; he also emphasizes its moral dimension. In movies we can get away with observing
without responsibility. André Bazin remarks in another context:
"Incontestably, there is in the pleasure derived from cinema and
novel a self-satisfaction, a concession to solitude, a sort of betrayal
of action by a refusal of social responsibility." In some of his movies

Hitchcock exploits this irresponsibility: "[In *Notorious*] the public was being given the great privilege of embracing Cary Grant and Ingrid Bergman together. It was a kind of temporary *ménage à trois.*" In a basically comic film like *Notorious* the audience can remain irresponsible, but in his best films the irresponsible audience must go through the punishment of terror. And Truffaut's approach breaks down most clearly when he is faced with what may be Hitchcock's most perfect expression of the interdependence of his themes and techniques—*Psycho*. In *Psycho* Hitchcock brings the voyeuristic assumptions of film form to the surface and in the process brings his audience from the detachment of irresponsible spectators to the involvement of implicated participants.

Hitchcock's films frequently approach the problem of detachment and involvement through separate but complementary treatments that might almost be called "genres." In "comedies" like *The Lady Vanishes, North by Northwest,* or *Torn Curtain,* the central characters are a romantic couple, with whom the audience automatically sympathizes. They serve as audience surrogates in a series of adventures that turn out happily. The axe is never far away from the neck in these comedies, but all conflict is finally dissipated by the end of the film, frequently by near fairy-tale or romance means. At the end of *North by Northwest* Cary Grant tries vainly to pull Eva Marie Saint to safety, while she dangles from the face of Mt. Rushmore. He can't do it. But then he can do it. The straining impossibility turns into fairy-tale ease. He pulls her up—into the top bunk of their Pullman, speeding away from the Dakotas.

Hitchcock's tragedies have no such romantic couple for ease of audience identification and sympathy; Truffaut remarks that there is no one in *Psycho* to identify with. We cast around without bearings, looking for conventional movie clues to tell us we have found the "right" character. But everyone is suspect. The first possible romantic couple in *Psycho*—Sam Loomis and Marion Crane (John Gavin and Janet Leigh)—have a melancholic relation in which sex and money are the prime topics of conversation. The later relation between Sam and Marion's sister Lila (Vera Miles), because it is founded on such dubious grounds, only emphasizes that *Psycho* is not the place to find a romantic couple. Solving a mystery may bring together Margaret Lockwood and Michael Redgrave in *The Lady Vanishes,* but it does not work in *Psycho*. Neither Sam, nor Marion, nor Lila, is particularly attractive. We can never give any of them our full sympathy, although we are often sympathetic to each. And Hitchcock manipulates our desire to sympathize and identify. He plays malevolently on the audience assumption that the character we sympathize with most, whose point of view we share, is the same character who is morally right in the story the movie tells. He gleefully defeats our expectation that our moral sympathies and our aesthetic sympathies remain fixed throughout the movie.

Hitchcock begins this manipulation at the very beginning of *Psycho*. He forces the audience, although we may not realize it immediately, to face the most sinister connotations of our audience role—our participation in the watching and observing that shades quickly into voyeurism. We see first a long view of a city and titles that read successively "Phoenix, Arizona. Friday, December the eleventh. Two forty-three P.M." We sit back and turn on the "objective" vision we reserve for documentaries, the aesthetic equivalent for a detached contemplation of the truth. But we are forced instead to watch an intensely personal, even embarrassing, scene. The camera moves closer and closer to one of the buildings, until finally it ducks under a drawn shade and emerges in a hotel room where Marion, in bra and halfslip, and Sam, bare to the waist, are having a late lunch-hour tryst. Perhaps we can call on our documentary detachment to insulate us from this scene, and thereby resist Hitchcock manipulations. Truffaut insulates himself by an interest in plot dynamics: "The sex angle was raised so that later on the audience would think that Anthony Perkins is merely a voyeur." But throughout *Psycho* Hitchcock continually assaults our claims of objectivity and detachment in order to emphasize and illustrate our real implication.

Hitchcock successively involves us with Marion and then Norman Bates (Perkins) through the gradually increasing use of a subjective camera. In both involvements there is at first a residual doubt, a nagging compunction about the moral aspects of our aesthetic involvement. In terms of conventional movie morality, or what our second-guessing has provisionally told us about the morality of *Psycho,* Sam and Marion are wrong; she's even overstayed her lunch hour. Hitchcock plays on our desire to feel superior because we have figured out *Psycho*'s system of rewards and punishments: "You know that the public always likes to be one jump ahead of the story; they like to feel they know what's coming next. So you deliberately play upon this fact to control their thoughts." He invites us next to feel morally superior as well as aesthetically. We can make a few moral distinctions on the basis of this first conversation between Sam and Marion. They can't get married and can't even find a pleasant place to meet because Sam has no money, at least not enough both to get married and to pay off his ex-wife's alimony. The lecherous rancher in Marion's office confirms our acceptance of the Sam-Marion relationship. What poetic justice it would be if his sexually tainted money could be used to make the dreams of Sam and Marion come true! By this point we have gone

beyond Marion. We wait impatiently as she moves about her bedroom, debating whether or not to take the money; through Hitchcock's manipulation of our moral responses, we have already decided.

Our identification with Marion becomes more directed as we
drive away from Phoenix with her. We sit in the driver's seat and
look out the window; when we look at Marion herself, we hear the
voices in her head, fantasies about what everyone in Phoenix must
be saying. Except for the single establishing shot in which we see
the police car pull up near Marion's parked car (and after all, at
this time she is asleep), we remain inside the car with her, limited
within the world of her imaginings, accomplices with her—for a
time—in what she has done. The state trooper appears as a
figure of vague malevolence; his shades reinforce his blankness.
When he waits across the street from the used car lot, we are appre-
hensive with Marion. When she drives away and an offscreen voice
yells "Hey!" we know it's the trooper. But it's not and he really
doesn't seem to be waiting for Marion at all. Through the subjective camera and the audience's belief in economy of means
("every character fits in somewhere"), Hitchcock has given us that
guilty, almost paranoid, state of mind that converts all outside
itself into images of potential evil.

This feeling of guilt begins to dissipate when we arrive at the
motel owned by Norman Bates and his mother. Norman is a genial,
shy young fellow, unassuming, pleasant. He's friendly, he makes
jokes, he even invites nervous Marion to dinner. When his mother
makes him withdraw the invitation, he talks to Marion feelingly
about the traps life has put him in. Marion callously suggests that
he should have his mother committed, "put someplace." We are
beginning to turn against Marion. Norman is a sensitive boy and
he loves his mother. Once again our conventional reactions come
into play. We wonder if we have been wrong about Marion. Perhaps she did have some cause for the theft, but she has a bad
streak. And that first image of sex in the afternoon may recur as
proof. She invites Norman into her room, but he draws back. Was
her sexuality a threat to Sam in the same way?

Hitchcock's gradual separation of our sympathies from Marion
and attachment of them to Norman now becomes even more
delicate. We follow Norman into the next room and watch as he
moves aside a picture to reveal a peephole into Marion's cabin.
He watches her undress and, in some important way, we feel the
temptress is more guilty than the Peeping Tom. In the first scene
of the movie Marion wore white bra and white halfslip. When she
finally decided to take the money, while it lay on her bed as she
packed, she wore a black bra and halfslip. She drove off in a black

car and then traded it in for a light-colored model. But our conventional moral-aesthetic sense can't be fooled. Once again, as Norman peers through the peephole, we see the black bra and halfslip, and remember Marion's guilt, a guilt we do not want to share. This perhaps dubious pattern of dark and light only reinforces something more basic. Whether we realize it or not, we have had a Norman-like perspective from the beginning of the movie. We too were Peeping Toms when we looked through the window of the hotel room Sam and Marion rented. We shared the Peeping-Tom exposure of Marion when her boss noticed her (and us) staring at him through the car window. When we look through the peephole with Norman, we are doing something we have done before; this time, like the first time, we know we won't be caught. We tend to blame Marion and not Norman because we are fellow-voyeurs with him, and we do not want to blame ourselves.[2]

It is worthwhile to emphasize the way Hitchcock manages our shift from Marion to Norman, since many commentators on *Psycho* assume that Marion's murder is somehow justified because she is a thief. But ironically enough her talk with Norman has convinced her that she has done wrong and should return to Phoenix. Her last act before the fatal shower is to figure out how to cover from her own bank account the loss sustained in buying the car. But her bra and halfslip have already given her away to Norman, whose psychotic view of people admits no shade between black and white, no difference between a mildly flirtatious invitation and a blatant proposition. Hitchcock masterfully implies that we can't tell the difference either. Perhaps the murder may also sardonically mirror our beliefs about Hollywood: Janet Leigh was the star of the first half of the movie; Perkins murders her and becomes the star of the second half. Perhaps we're also being invited to remember that Janet Leigh had recently disported herself sexually in another motel in Welles' *Touch of Evil* (1958; *Psycho,* 1960). In any case, Norman had added her to his collection of dead birds; when he emerges from the bathroom after his "first" look at her, he knocks one of the bird pictures from the wall. Marion fits well into the collection because, after all, her last name is Crane and she comes from Phoenix. But she won't rise again. There's only one phoenix, and in this movie it's Norman's mother.

The sight of Norman cleaning up the bathroom after the murder reinforces our identification with him aesthetically and morally. Our hands hold the mop and swirl the towel around the floor;

---

[2] Because of the importance of the motif of observation, especially through windows, it's worth noting that we see Hitchcock through the window of Marion's office.

Hitchcock cryptically remarks to Truffaut about his own hyper-cleanliness. Norman cleans up so well because he is a dutiful son trying to protect his crazy mother. Once again, Hitchcock forces us into the security of conventional moral reactions in the face of an absurd situation. In many of his movies he begins with an excessively normal, even banal, situation and then proceeds to show the maniacal forces seething just below the surface. Norman's mop reverses the process; the bathroom is gleaming and conventional once more. We are relieved that the most characterless place on the American landscape has become characterless once again. We have become so identified with Norman's point of view that we feel a moment of apprehension when the car refuses to sink all the way into the black pool. But it finally goes down. We heave a sigh of relief with Norman; the insanity has been submerged once again. Our relief masks our progress from the acceptance of illicit sex to robbery, to murder, what Truffaut with his rage for precision calls a "scale of the abnormal." The memory of our pleasure in Marion's nudity, even while the murder was in process, our effort to see if that was a breast or only an arm we half-glimpsed, all become submerged, especially since, with Norman, we may have decided that she deserved it.

Our sympathy with Norman also controls our feeling about the detective, Arbogast (Martin Balsam). Arbogast upsets Norman with his questions, and we have little or no sympathy with him through the camera. When he walks upstairs in the house, we get only one short shot of his lower legs. Then all the shots are face on, as if we were at the top of the stairs with "Mother." When the murder begins we look straight into Arbogast's face as he staggers back down the stairs under the knife blows. We follow him along with "Mother," striking again and again. The conventional and self-protective operations of our aesthetic and moral sympathies have once again implicated us in something we were not ready for. Hitchcock plays to Truffaut's prejudices by saying that the high camera shot—the bird's-eye view—that begins the murder segment was used to avoid showing "Mother's" face. But when he returns to it at the end of the scene, as Perkins carries her down to the fruit cellar, Hitchcock checks off our complicity. We are no longer so terrified.

Sam and Lila arrive during the day, presaging the illumination of Norman's dark subconscious. Previously the dark brooding vertical shaft of the house had stood high in the shadows behind the banal well-lit horizontal of the motel. With light now striking them both, the house is potentially no longer so mysterious. Sam cannot go in to discover the secret. Like Marion and Arbogast, he had first visited the motel (in one of the few inept scenes) at night.

But this is Lila's first visit; Sam delays Norman through conversation. His bad acting (on two levels) and accusations of Norman keep us sympathetic to Norman and divided from Sam. In the house Lila has begun to move through the rooms and examine the furniture of Norman's mind. She sees a movement behind her and turns to find a full-length mirror. Like the audience, she has rummaged around in someone else's inner darkness and discovers there, instead of unknown horrors, something akin to herself. With Norman's return she races toward the fruit cellar and the final secret is revealed—"the foul rag-and-bone shop of the heart."

Norman's psychosis is the MacGuffin of *Psycho*; its special nature is irrelevant. Hitchcock concentrates instead on problems of presentation and point of view, the uncertain line between the normal audience and the psychotic character, and the actually hazy areas of moral judgment. Throughout the movie we are placed in situations that challenge our conventionalized aesthetic and moral responses. Hitchcock's attack on the reflex use of conventional pieties is basically an attack on the desire of the audience to deny responsibility and assert complete detachment. The viewer who wants such placidity and irresponsibility is mocked by the pseudodocumentary beginning of the movie. If he chooses, he has another trapdoor available at the end—in the explanation of the psychologist.

Because Norman has murdered both his mother and her lover, we don't have the conventional out of psychiatric exoneration from guilt. But the psychologist does offer us a way to escape responsibility by even more acceptable means: He sets up a screen of jargon to "explain" Norman. For the viewer who has learned anything from *Psycho* he must be dismissed. The visual clues are all present: He is greasy and all-knowing; he lectures and gestures with false expansiveness. But it is his explanations that are really insufficient. And one wonders if any categories would be sufficient. Like the moral tags dispensed by the Chorus at the end of *Oedipus Tyrannos,* the bland wisdom of the psychologist bears little relation to the complex human reality that has been our experience in the rest of the movie. We understand Norman because we realize the continuum between his actions and our own. We leave the front office of "clear" explanation, while the psychologist is still talking, to enter Norman's cell. Through Hitchcock's manipulation of point of view and moral sympathy, we have entered the shell of his personality and discovered the rooted violence and perverse sexuality that may be in our own natures. Our desire to save Norman is a desire to save ourselves. But we have been walled off from the comfortable and reasonable and "technical" explanations of the psychologist. The impact that *Psycho* has upon us shows how deeply we've been implicated.

In 1955 Truffaut and Claude Chabrol had gone to interview Hitchcock on the location set of *To Catch a Thief* at Joinville. In their excitement they walked on the ice of a little pond in the center of a courtyard and fell in, tape recorder and all. Truffaut turns this into a charming anecdote: "It all began when we broke the ice." But he conducts the interview as if this first encounter were cautionary. It symbolizes his unwillingness to leave the surface and plunge, however uncertainly, into the dark and icy depths.

Inside Norman Bates

RAYMOND DURGNAT

◆◇◆

The camera climbs towards a window like any other window. Documentary-style, a subtitle states time and date; but it really means: Here and Now, at this moment, without warning, imperceptibly, destiny entered these lives. On a hot day, during their lunch-break, in an impersonal hotel bedroom, Marion Crane (Janet Leigh) and Sam Loomis (John Gavin) are half-naked and necking. The nightmare begins at noon. The heat, the bleached feel of the visuals, the half-nakedness, the time, evoke an atmosphere of unsatiated sensuality (indeed, the heavy petting of so many of Hitchcock's American films, from *Notorious* to *North by Northwest,* suggests a frustrating coldness, even, intercourse with neither orgasm nor emotional relief). In a very matter-of-fact way the lovers are discussing the man's divorce and the money they need if they are to marry. The general situation—half-stripping at lunchtime and then talking about cash—is vaguely offensive; yet they seem decent people, we accept and care about them. This ambiguity pervades their whole relationship. In some way Sam seems petulant, weak, unworthy; in others, Marion seems prim, tough, less concerned with unconditional love than with—respectability? Are they in love or only convinced they are? At any rate, we're not especially

From Films and Feelings *by Raymond Durgnat (Boston: M.I.T. Press, 1967), pp. 209–20. Copyright © 1967 by Raymond Durgnat. Reprinted by permission of the M.I.T. Press and Faber and Faber Ltd.*

anxious for them to get married. In default of the money, she is tempted to break off the affair, and we are sufficiently disquietened to watch with something between curiosity and concern, rather with an eagerness for them to get married and live "happily ever after."

Marion returns to the sane, shallow, superficial people of the office where she works. It's not long before sex and cash are intertwined again. A fat client makes a rather coarse and vulgar attempt to flirt with her, brandishing a fat bankroll in her face. The other office girl, a plain and silly creature, is naïvely jealous of these gross attentions. "I expect he saw my wedding ring." Her self-consoling remark rubs salt in Marion's wound. We agree with her feeling that she is too pretty, efficient, sincere in love, to deserve to be worse off than this other girl. The fat customer brags that he wouldn't miss the money if it were stolen, and Marion's boss absolutely insists on entrusting it to her. Such smug, imperceptive responses all round reinforce our feeling that Marion has as much right to this excess money as its actual owner. These pinpricks accumulate into a kind of obsession and reinforce the confusion between her respectability (or pride) and her love (or sensuality). The money seems to offer a solution to all these "raw edges" of feeling. Her theft is (so to speak) an impulse born of converging obsessions, which suddenly click into place forming an irresistible urge. It is also a tribute to her daring, her strength of passion; there is an element of moral *hubris vis-à-vis*. There is also an element of *hubris vis-à-vis* her lover, as if in acting so boldly where he has been so weak she is taking over the initiative—and is not going to be thanked for her devotion. Soon she is driving hard away from the town, tormented not so much by conscience as by fear. We can't believe she'll get away with it, especially as criminals never do in American films. We hope she will, and there is still a get-out: The theft won't be noticed until Monday morning, she can always return the money. Will she go on to decide to return it, but lose it? Will someone else steal it from her? Will Sam betray her, by his weakness, somehow?

A big, brutal-looking motorbike cop with dark glasses trails her, suspiciously. His menacing figure recalls the lawbreakers of *The Wild One* and the motorcyclists of *Orphée* who ran men down in the name, not of justice, but of a law above the law, the brutal Will of destiny. He is "the law," but he has a special, *personal* brutality of his own. Is he really following her, or is she only imagining he is? The psychological pressures complicate and intensify. To shake him off, she exchanges her car at a garage run by a very obliging character, apparently the very antithesis of the cop. The cop is saying, "I remind you of punishment. Turn back!," the garage-hand, "I make crime pleasant and easy, go on." She acquires a white car—the color of her underwear in the necking scene, the color of innocence and dissatisfied sensuality; but all her precautions are of no avail. The cop still tails her, a terrifying dark angel sent to give her a last chance. Or sent simply to torture her, to diminish her chances; for without him she has a weekend in which to repent. There is danger of, as it were, rape-by-justice. We sigh with relief when at last she shakes him off.

She is beyond the reach of the law—or fear—now. But—where is she? The rain pours down across the windscreen, blurring lights and creating a wavering landscape. She is in what in *Orphée* is called *la Zone,* the no-man's-land between reality and the nightmare. The cop was both danger and safety. It is almost as if he were sent, after all, not to turn her back, but to make her drive on. The theological notion of double predestination provides a clue, "God sends sinners a chance to repent *in order that* by rejecting it, as he knows they will, they will damn themselves more thoroughly than ever." But as she reasons with herself, she is beginning to realize the futility of her theft—Sam is too sensible to accept the money. . . .

The rain forces her into a motel, managed by Norman Bates (Anthony Perkins). Norman is an engagingly naïve country youth, very honest, unconcerned with making money, almost a symbol of rustic virtue and country contentment. The whole film hinges on his sensitivity and charm—we tend to like him whatever his faults. His friendliness is all the more reassuring in contrast with the sinister atmosphere (the stuffed birds of prey, the Victorian house just behind the motel, where his petulant, tyrannical old mother lives). He seems tainted by the atmosphere, but the over-obvious horror clichés shift our suspicions from Norman to the atmosphere; they camouflage the inevitably stilted presentation of his relationships with Mrs. Bates; they contrast with the slick, modern, informal style of the film as a whole. Mrs. Bates comes from Norman's childhood and it's fitting that she should exist in an aesthetic idiom now considered childish—she would feel quite at home in James Whale's *The Old Dark House.*

Marion calls Norman's bird-stuffing a rather morbid hobby and says Norman resembles the dead birds of prey. Hitchcock plays fair with his audience, even while misleading us. True, he lets us believe in Mrs. Bates—but so do Marion and Norman. Maybe, as the psychiatrist says later, Norman was never entirely Norman, he faintly knew the truth about Mrs. Bates—but then again Mrs. Bates is very stilted, we only half-believe in her.

Norman cheerfully admits to his faults of character; he is a very reasonable, modest guy. Gradually Marion realizes that she is his superior, that, if unhappy, she is self-possessed, whereas his "contented" acquiescence in looking after his domineering mother has something weak and helpless. His wisdom about money and the example of his servitude help to free her from the power of her impulse. She realizes that what she stole was not love but only money, an attempt to avoid her problems. Norman is almost a sacrificial victim whose tragic example frees her.

But he is not a hopeless case. We feel that she owes it to him to return the favor. We want him to be freed from his horrible mother, for he is a decent fellow. There is something dissatisfying in Marion's decision simply to return, alone, to the everyday, with its little degradations, its mutually exclusive choices—while leaving Norman here, unhelped. A sort of bewilderment percolates through the audience at this weird, premature "happy ending." We are, so to speak, in another "zone."

The film elaborately establishes Marion's search for a hiding-place for her cash. The search seems to turn her indifference to Norman into an entrenched cynicism, for he isn't the sort of lad to steal it. As she undresses, Norman watches through the peep-hole. We laugh very uneasily at his avid voyeurism, but it does not quite put him in our bad books, for he has been lonely and dominated by his puritanical mother and his spying on Marion represents a movement towards normality and freedom, which we want for his sake. This is almost a dissatisfying love scene (like necking for lunch). The erotic overtones are juicy, and please us. And we are pleased to feel the story moving again.

The "movement towards" Marion is intensified—with a vengeance—when Mrs. Bates with a knife upraised charges in and stabs her to death in the shower. The murder is too erotic not to enjoy, but too grisly to enjoy. Its ferocity and pornography are opposed, we are shocked into violent protest and horror, yet they force on the average spectator a rapid, hysteric, moral oscillation between protest and enjoyment. There is a Hays Code sort of moral in the air: "Look what thieving necking girls get," but her fate is also ironically unjust, for she had just resolved to return the money.

If the Peeping Tom episode is a "weak" yet eerie version of the hotel scene, the murder is a sarcastic exaggeration of it—her sensuality's satisfied now, all right. We feel guilty about enjoying this film, but we have to admit we're having our money's worth of fun and fear.

Mom would be a convenient scapegoat; but we are headed away from complacent hatred back into something subtler and far more uncomfortable by Norman's distress at her crime and his concern for her. In the next sequence, he begins mopping-up operations in the bathroom, the action of an exceptionally dutiful son. The presence of Marion's naked corpse is both erotic and extremely uncomfortable. The film offers us a "first-person" experience answering the question which so often occurs to crime fans, "Would I be able to get down to the practical details of clearing up the corpse and the blood"—a thought which appals many people more than that of the actual killing. The answer the film gives is, "A sensitive and dutiful son like Norman can—therefore, so could you, if you really had to." We watch Norman doing it, and the feeling that we could too is gratifying to the worse side of our nature, but upsets the other.

Although there is a quietly disturbing contrast between Norman's usual sensibility and his matter-of-fact practicality on this particular chore, we feel that in his way he was on the edge of being "liberated" by his interest in Marion, that she slew Marion so as to keep him, and that in covering up for Mom, Norman is turning the other cheek, manifesting the equanimity and charity of a saint. The spectator's moral purity is being outflanked at both ends—by morbid, pornographic interest and by a sympathetic pity for charming Norman.

Not that indignation and disgust are lulled asleep. On the contrary. For example, there is a very precise mix between a close-up of the plughole down which our saintly voyeur is swabbing the blood and a close-up of Marion's open eye staring at us as if to say, "What about my feelings? Why don't you interview the dead?" She's peeping back at us from beyond the grave, from down the drain, with protest and indignation, eternal and colossal—or surprise and fear—or just nothing. This visual rhyme is not just a piece of sadistic wit, but a little essay in metaphor; it never does to interpret visual effects too definitely, but, e.g., the plughole is like an eye socket, the eye ("Window of the soul," as they say) is just a mushroom out of a black hole. There is a sense of total nothingness and if the "joke" provides a little hysteria which relieves the horror faintly, it insinuates a subtler unease: We must be mad to be laughing at a joke like this.

Norman chews candy as he watches the white car sink beneath the very black surface of the swamp behind the house. As the film uses psychoanalytical ideas, it's appropriate to use them on the film—the bathroom scene, very glossy and white, and devoted to the theme of cleanliness, is followed by a scene in which everything disappears into a black sticky cesspool. Norman has pulled the chain.

When the car sticks instead of sinking, we are alarmed, but when at last it disappears we heave a sigh of relief. Thank goodness! Norman is a good boy (despite the candy), it would be wrong to punish him, Marion's a corpse, it's no use crying over spilt blood, bury her quick, tidy up, get her out of the way! But when Norman

tosses in the thick wad of cash, which he thinks is just an old newspaper, a cry of shock and regret is wrested from the audience. That
valuable money, what a waste! Norman's saintly indifference to
Mammon hurts us. We want to forget Marion probably because her
murder shook us up so much. But the money had become "what
she died for, what she hid," that is, virtually a substitute identity.
Its derisive disappearance creates hysteria as again the narrative
seems to "end."

Sam Loomis discusses Marion's disappearance with her sister Lila
(Vera Miles). The visuals are grey and scruffy. The setting is Sam's
wife's ironmongery store where callous chit-chat about insecticides
is overheard and pitchfork prongs are visually prominent. The drab
everyday is full of trivial or latent cruelty. The meeting of lover
and sister is hostile, but their disputes are ironically complacent
compared with the terrible truth. Lila seems more sensible, more
adult than Marion, and perhaps more righteous—but also worried,
subdued. A private detective, Arbogast (Martin Balsam) insists on
introducing himself, and tells them that Marion has absconded
with the money. They refuse to believe him. They detest his coarse,
obnoxious approach—so do we, and, like Sam and Lila, feel he
must be up to some dirty game. His cynicism doesn't fit Marion's
case—although, in a sense, it is justified.

As he tracks Marion down to Norman's mansion we half want
him to fail—for Norman's sake, and because he may be up to some
cynical scheme of his own. . . . Just before he confronts Norman
we realize that he is completely, admirably honest. In the battle of
wits between Norman and Arbogast we sympathize with them both
—Marion *must* be avenged, Arbogast is tough enough to uncover
the truth; and yet Norman's motives are selfless, and perhaps Mrs.
Bates will be more than even Arbogast bargains for. As he climbs
the stairs towards the old lady's room, we realize clearly that his
pushful cynicism, hitherto his strength, is now his weakness. He is
formidable, and physically is probably Mom's match, but he is too
naïve to be looking for whatever he'll find—and Mrs. Bates comes
tearing out of her room with the superspeed of the superstrong
insane and with repeated jabs of her knife sends him tumbling
backwards down the stairs, dead, just like that. Is Mom invincible?

Another car sinks into the swamp, the narrative "ends" at another nihilistic moment.

The whole plot, which has twice ended so disastrously, starts
again, as Sam and Lila come to investigate the disappearance of
the investigator who came to investigate the disappearance of. . . .
Probably by now most spectators have guessed that Mom=Norman.
But we can't be sure, in such a film. The only thing we can be

certain of is the imminence of violent death—again. What matters is not whether we know, but whether Sam and Lila find out—or get killed. They might. Heroes and heroines do, in this film. And if they do find out, what will happen to Norman—saintly accomplice of two—at least—crimes . . . ?

The determined, but prosaic and therefore perilously naive, couple call on the local sheriff (John McIntire) who explains that Norman is eccentric but harmless, that Mom has been dead and buried these ten years past, and so on. But we heard Norman persuade Mom to hide in the cellar and we saw Mom come tearing out of her room to kill Arbogast. The sheriff's clue is so wrapped up in complacency and ignorance that instead of clarifying our suspicions it confounds them further. The sheriff's suggestion opens up astounding new avenues of depravity: "If Norman's mother is still alive, then who's the woman buried up there in Green Lawns Cemetery?" [1] If they believe what the sheriff says, they will never go to the old house, and then how can Marion and Arbogast be avenged? But *if* they go there. . . .

Sam keeps Norman talking while Lila sneaks into the house to explore; clearly the most dangerous game to play, especially with a possible Mom waiting for her. As we can't make up our mind whether the danger is coming from in front of her (Mom) or from behind her (Norman), we're no longer thinking very coherently, and as we can't make up our mind what we want to happen to Norman, we yield to a helpless hysteria.

Norman grows more anxious and angry as Sam brutally presses him; he struggles to keep his temper, to quieten his tormentor's suspicions, while keeping Mom from breaking out in himself (if you know) or (if you don't) bravely protecting his Mom or (if you're not sure) both or neither or which? The scene almost shifts our sympathies round—such is Norman's sincerity—to: "brutal smug adulterer bullies sensitive kid into despair." After all, whether Norman is weak or maniac or both, he probably believes in Mom, he is only trying to obviate another climax, another killing, he is frantically on the side of peace.

Lila explores the house. Amidst the tension there is an unexpected intellectual interest, and pathos. Norman's rooms are a picture of his mind and everyday life. There is the record-player with the classical L.P. (so out of place in this Gothicy house), there are the fluffy childhood toys which are presumably still played with. Norman is weaker-minded, more sensitive, than we thought, which makes him more pathetic (and more surprising—menacing?).

[1] Well might he ask.

Norman, mad with suspicion, rushes from the motel into the house as Lila takes refuge in the cellar—where, we know, Norman puts Mom in times of stress. And Mom does exist, there she is, horribly old, evil and withered, at a closer look she's dead and withered, but still grinning malevolently, she's a ghost, and when Lila turns, there's *another* Mom, grinning malevolently, very much alive, knife upraised. There aren't no Moms, there are two Moms, then the second disintegrates, the wig slides off, it's Norman. It's not simply the surprise that shocks; it's the intensity of terror and the obscenity of the disintegration. In rather the same way, when Mom came tearing out of her room at Arbogast, she had the notoriously terrible strength of the insane, and a visible virility quite obscene in an old lady; the explanation doesn't explain *that* away; it intensifies its impact because illusion and explanation coexist.

We are relieved to hear that everything is going to be comfortably explained for us by the police psychologist (Simon Oakland). As soon as we see him we begin to dislike his brash, callous, know-all manner; he puts our backs up as Arbogast did. We expect the clichés: poor mixed-up kid, it was all the fault of stern, possessive, puritanical Mom. But gradually we realize he's not saying this at all. It was Norman who was jealous, who imagined that his (for all we know) normal Mom was a promiscuous Mom and murdered and embalmed her and then imagined she was a jealous puritanical Mom and then lived out two false characters— nice normal Norman and nasty Mom. So much for rustic contentment. Norman was never, we gather, entirely Norman, i.e., even when he was being charming and we felt sorry for him, he knew deep down what he was doing. The psychologist's explanation takes away our explanation: What we thought was "deep," the "solution," is merely the topmost level of nastiness. He restores terror, guilt, injustice. Up till now Mom's gruesome appearance has been in accord with her character: "Well, if she's dead, she asked for it, look at how she messed up her tender and devoted son." Now all this is reversed, the coconut-faced corpse was once a sunny, apple-cheeked mother. The boy has literally turned her into his fantasy of her.

But if the psychologist, brutal and cynical, is the most intimate of private eyes, the joker is still to come. All we've had has been an intellectual, rational explanation. Now we see Norman sitting against a blank, white, hygienic wall. He is in full-face close-up, his madness is rammed into the cinema. Briefly our entire world is his face, the thoughts behind it, *his* world. We have little else with which to identify. An utter flatness, whiteness, simplicity, in short, eternity. He is cackling to himself, in Mummy's mummy's voice.

She is jubilant because she is outwitting them all, pretending to be a sweet old lady who won't even hurt a fly. Mom has just killed Norman and disguised himself as him.

The Chinese sage wrote: "Now I do not know whether I was then a man dreaming I was a butterfly or whether I am now a butterfly dreaming I am a man." With Norman it's flies. His ricocheting self-punishment is so total that—well, we can hardly pity him, for there's no one left there to pity. And he or she or it seems to think it is escaping punishment, which is very immoral of him or her or it; but a nausea-*like* compassion makes itself felt. We are too thoroughly satisfied to hate.

The appearance of Mom's face under the madman's, and then of a skull under Mom's, has a climatic brutality, but also simplifies, liberates us from the baffling maze of malevolent Nothings which our sensitive boy has become. Needless to say, it is a simplification on the most nihilistic level: Are any of us realler than our skulls? There follows a shot of the police lifting Marion's car, wrapped in chains, from the swamp. There is no "decent obscurity." And Nothing to the nth degree has killed real people whom we sympathized with. But we too hoped the car would sink (just as we hoped Marion would get away with the cash). We too have been accomplices after the acts—futile acts.

People leave the cinema, chuckling incredulously, groggy, exhilarated yet hysterical, half ready to believe that everybody in the world is as mad as Norman. A kathartic indulgence in pornographic murder is succeeded by an embarrassed humility, an unsentimental compassion towards insanity. The entire film is a prolonged practical joke in the worst of taste. If it weren't in bad taste, it would not be kathartic, embarrassing or compassionate.

It is not just a sick joke, it is also a very sad joke. Because it is outrageous, it exhilarates, but it is a very depressed film as well. The by-play with the money is strange and disturbing. It is produced as a weapon of seduction by a repulsive but normal male. Its victim resents the implied insult but yields to the money. The money, she felt, would enable her to find, all at once, respectability, sensuality, love. It becomes the last clue, a substitute identity, an antisoul. Marion who hoped to avoid choice, and sacrifice (the *hubris* of American optimism), is reduced to a nude body, a car, a bankroll.

Everything piles up in the swamp—and is dredged up again. The film is not just a sick joke and a very sad joke, but a lavatory joke. It is a derisive misuse of the key-images of "the American way of life": Momism (but it blames son), cash (and rural virtue), necking (and respectability), plumbing and smart cars. The reality to which

Sam and Lila return is not a joyous one, but a drab shop of insecticides, pitchforks, and—in addition—a vision of horror. The plot inevitably arouses in the spectator a feeling that Lila and Sam could eventually, possibly, consolingly, fall in love. But there is no hint of it in the final image. Each is still alone. This is the sanity that balances the diabolical nothing which is the human soul. Marion, striving for everything, lost everything. Only Norman has defied society and superficiality and found "rest." Only Norman has found himself, and lost himself.

Like many films, *Psycho*'s aesthetic method is not that of providing enlightening information about its characters; it provides just enough to confuse us; it works by luring the audience into becoming the characters, sharing and living out their experiences within them in carefully determined patterns. The characters tend to be alone on the screen. Even the conversations are filmed mainly in alternating close-ups. The close-up both enlarges (intensifies) and isolates (blots out the rest of the world). While each character is speaking the spectator sees, feels, becomes him and only him. The next shot wrenches him into becoming the *antagonistic* character. Our sympathies alternate rapidly—our feelings are poured into so many molds which are distended or smashed by contradictions, revelations, twists. Simple as the characters are, in principle, they are, because well acted, convincingly real. The atmosphere is hypnotic, the events so outrageous and managed with such brinkmanship of taste, the hints, allusions and subversive shifts of sympathy are managed with such sly tact, its constant emotional collisions are so quick, subtle and drastic, that the "sketchiness" of the characters no more invalidates them than it invalidates the plays of Racine.

In its powerful vagueness, it works on the spectator not unlike music. It is planned, felt out, in terms of varied motifs, of emotional chords and dissonances, of patterns. Hitchcock has a very refined sense of sly or brash emotional discords, of how to modulate and combine them. The coarse customer, the cop, Arbogast, and the psychologist are incarnations of the same force—unpleasant common sense. The trusting boss, the garageist, the local sheriff, Norman himself all agreeably further evil. The woman in the ironmonger's who is determined to kill insects painlessly is mirrored in Norman's final crone-voiced cackle that he won't even hurt a fly (Is absurd squeamishness the hypocritical form of homicidal mania?). Norman, in conversation, unwittingly frees Marion of the compulsive theft which Sam inspired in her. But Sam bullies Norman like the cop bullied Marion.

Lila is a more reasonable, but "joyless" double of Marion. Sam loses Marion to Norman but Norman is destroyed by Sam and

Lila. Lila, in a sense, is Marion "come back"—a parallel to the "second Mom" in the basement. As Lila roams through Norman's rooms, she is almost the substitute mother, the young woman who is kind and normal and will therefore destroy him. Norman and Sam are both dark-haired, faintly resemble each other. Norman killed his mother because he thought she had a lover; and is destroyed by a young adulterer and his mistress's sister. The three penetrations to "the truth about Norman"—Marion's, Arbogast's, the young couple's—are like three movements in music—the first two themes are contrasted (a sensual theme involving a girl, an unromantic theme involving Arbogast, the third combining them —a young couple who aren't quite romantically connected).

All these patterns, like inversions of certain emotional chords, result from the film's simplicity of form, but they are like haunting harmonies placed on a simple, yet eerie melodic line. The cutting has a quick, ragged, Stravinskyian rhythm.

The minor quirks and sins (adultery, a "thing" about insecticides) of the normal world are the tips of the horns of the real reality, concealed beyond, or below, the "zone." In *Psycho* nothing that isn't disturbing or tainted ever happens, and to enjoy it (as most people do) is to stand convicted, and consciously convicted, of a lurking nostalgia for evil (i.e., of thoroughly enjoying it in fantasy). Norman's big mistake is that he let his fantasies enjoy him. The film is a practical joke: It convicts all the spectators of Original Sin. One does not so much watch, as participate in, it, as one might in a religious ritual involving the confession and a— well, one cannot say that absolution is granted. On the contrary, we have to take what comfort, or discomfort, we can from the implied complicity.

Hitchcock may have had a Jesuit education, but surely *Psycho* isn't a Christian film; it has a Dionysiac force and ruthlessness; one might call it a Greek tragicomedy.

Macabre Merriment
JOHN CROSBY

❖❖

The best thing about "Alfred Hitchcock Presents" is Alfred Hitchcock presenting. Mr. Hitchcock, as I suppose every one now knows, is a man of redoubtable circumference. In fact, he might be described as one of those Englishmen on whom the sun never sets. And he has a face like a slightly malevolent kewpie doll.

On his program, the physiognomy and body are first outlined gently in pencil, then inked in. Finally, the man himself floats in, still in darkness. Then they turn the lights on and there he is. CBS, I guess, feels that it would be too great a shock to present the man, cold, without advance preparation, and maybe they're right.

Mr. Hitchcock then purports to give us some advance warning about the story we're about to hear. Actually, he does no such thing, his introductions being magnificently irrelevant. On his last program, for instance, he held up a book: "I've been reading a mystery story. These paperback books will never replace hard covers. They're just good for reading. This one is full of the sort of advice mother used to give: Walk softly and carry a big stick. Strike first!"

This had nothing at all to do with the story that followed. It was a wonderfully macabre little tale about a business tycoon, played by Joseph Cotten, whose car is smashed by a bulldozer, operated by a road gang of convicts. The smashup totally paralyzes Cotten, and he has to lie there in the wreckage while the convicts steal his tires and his clothes. Finally, rescuers arrive, but they think he's dead, too, and bundle him off to the morgue.

Meanwhile, Cotten's mind is fully active, and his thoughts, expressed in narration, are hardly pleasant. Grim, isn't it? Of course, one thing about Hitchcock's handling of the macabre is a sort of sunny matter-of-fact quality which robs it of a lot of its sting. He just assumes that bodies lying around don't bother people especially, and presently they don't. As a matter of fact, inert bodies hold great fascination, for Hitchcock at the moment. His last picture, *The Trouble with Harry*, was about another inert body. This one was quite dead, but it had to be played by a live actor (Philip Truex). It must be restful for the actors—just lying there. Years ago, when the deadpan school of acting first gripped Hollywood, I had a terrible premonition that acting would some day come to this, and it has.

After the story is over, the Hitchcock visage appears again, to sign off, as it were, and these little messages are also worth listening to. One sign-off went like this: "And as the cold New England sun sinks behind the coroner's office, we take our leave of New Hampshire and prepare ourselves for the chant of the sponsor's message. Then I'll float back." After the chant of the sponsor's message, he floated back to say: "You may now leave your seats, chat with friends, have a smoke—but remember to come back after the intermission." The intermission, he added casually, would be one week long.

This one, starring John Forsythe, was about a man returning to his home town to track down his father's murderer, only to discover that he's the guy who did it, a tricky ending I have no intention of explaining. I throw it in only as a clue to the sort of fare Hitchcock dishes out week after week. I find them quite entertaining, very well acted and produced.

When he went into television, Hitchcock issued a communique which ran in part as follows: "Several people, addicted to sadistic reflection, have quite pointedly asked me how, having been accustomed to working months in filming a motion picture, I expected to commit a coherent story to film in a mere matter of two days. The very idea of Alfred Hitchcock, the calm, complacent Hitchcock, beset by the frantic frenzy commonly associated with television seems to have provided many hilarious moments of contemplative amusement to those who have less important things with which to concern their thoughts. It annoys me, this notion that I cannot move around rapidly when the occasion demands." So far he's managed to move that large frame around rapidly enough.

Hitchcock's TV Films

JACK EDMUND NOLAN

❖◇❖

Of the 353 [1] hour and half-hour TV series films produced under Alfred Hitchcock's aegis 1955–65, only a minority were directed by Hitchcock himself. Of these, Hitch has said to various reviewers that he didn't select properties for personal attention, but directed a few segments as they came up on the floor at Shamley Productions. Such statements are belied by the Hitch-directed segments themselves; they tend to utilize his favorite stars from his feature films (as well as favorite scripters and more season premiers directed by the "master of suspense"). Moreover, typical Hitchcock thematic material (the exchange of guilt, friendship between disparate types, the woman at bay) obtrudes in those segments which he directed.

The index which follows therefore attempts to fill a gaping historical "hole." To date, only three critics (Peter Bogdanovich, Jerry Vermilye, John Russell Taylor) have attempted to list Hitch-directed TV shows; the work of all these three includes omitted segments and date shufflings. None of them has tried to provide analyses of favorite Hitch themes as they were reflected in the director's TV work.

This is understandable, since research in U.S. TV films is a chaotic business. There are *no* reliable guides to dates, technical credits, or plots. The daily *New York Times* listing is the best source for TV data, but naturally only applies to the New York TV market area (one of 70 such areas). *TV Guide* remains a good source, but all its market-area editions must be consulted. In addition, particularly during the '50s but still true today, TV segments occasionally

From Film Fan Monthly, *June 1968. Copyright © 1968 by* Film Fan Monthly. *Reprinted by permission of the publisher.*

[1] Evidence indicates twelve additional half-hour segments were shot but not released first-run with the *Alfred Hitchcock Presents* show—They were not televised until after Spring 1965 when the various Hitchcock shows were placed in syndication/rerun.

premiere in, say, Toledo and are not seen in New York or Los Angeles until months later.

At least, the three writers above *tried* to fill a gap. Others have analyzed Hitch with little or no reference to the twelve total hours of filmed, fictional TV work he directed. Even Truffaut's magnificent book-length interview with the director omits all reference to his TV films from its otherwise perfect index and only includes a couple of references to them in the body of the text.

In the following index, AHH: Alfred Hitchcock Presents, *hour* version; AHP: Alfred Hitchcock Presents, *half-hour* version; FST: Ford Star Time; and S: Suspicion.

1. *Breakdown,* AHP. CBS, 13 Nov. '55. Original script by Francis Cockrell and Louis Pollock. Joseph Cotten. A tour de force by Cotten, who plays an accident victim thought to be dead. Paralyzed and finally placed on a mortuary slab, a tear from his eye signals an attendant that he's alive. Shot before *Revenge* (below) but televised after it.

2. *Revenge,* AHP. CBS, 2 Oct. '55. Script by Francis Cockrell and A. I. Bezzerides from Samuel Blas' story. Ralph Meeker, Vera Miles, Frances Bavier. An aircraft worker's wife (Miles) tells hubby a man attacked her; after hubby clubs the supposed culprit to death, he finds that his crazed wife thinks *many* men are guilty of having attacked her.

3. *The Case of Mr. Pelham,* AHP. CBS, 4 Dec. '55. Script by Francis Cockrell from Anthony Armstrong's story. Tom Ewell, Raymond Bailey, Kirby Smith, Kay Stewart, Jan Arvan, Norman Willis. The strangest property on which Hitch worked in any medium; Pelham (Ewell) believes that a duplicate of himself is taking over his life. A completely plotless item, it has Pelham go mad after his own butler accepts the double as himself.

4. *Back for Christmas,* AHP. CBS, 4 Mar. '56. Script by Francis Cockrell from John Collier's short story of the same title. John Williams, Isobel Elsom, A. E. Gould-Porter. Herb (Williams) kills his wife, buries her in the cellar and heads for the West Coast, U.S.A. In a typical Hitch twist finish, the placid Herb then receives a bill for repairs to his cellar (i.e., the body will soon be found), paid for by his wife as a present to him, before he killed her.

5. *Wet Saturday,* AHP. CBS, 30 Sept. '56. Script by Marian Cockrell from John Collier's short story of the same title. Sir Cedric Hardwicke, John Williams, Tina Purdom, Harry Barday. Upperclass Millicent (Purdom) murders her two-timing boy friend, and her dad (Hardwicke) can't hush it up.

6. *Mr. Blanchard's Secret,* AHP. CBS, 23 Dec. '56. Script by Sarett

Rudley from Emily Neff's story. Mary Scott, Robert Horton, Dayton Lummis, Meg Mundy. A bit of a switch on Hitchcock's feature *Rear Window*; a middle-class housewife (Scott) believes that neighbor Blanchard (Lummis) murdered his wife.

7. *One More Mile to Go,* AHP. CBS, 7 Apr. '57. Script by James P. Cavanaugh from F. J. Smith's story. David Wayne, Steve Brodie, Louise Larabee. This one includes an odd opening a la *Psycho*: Sam (Wayne) is shook up when a State Trooper stops his car on the highway, because Sam's wife's corpse is in the car trunk.

8. *Four O'Clock,* S. NBC, 30 Sept. '57. Script by Francis Cockrell from Cornell Woolrich's story. E. G. Marshall, Nancy Kelly, Richard Long. Hitch's first hour-length tv film, and a classic of suspense. Paul, a psychotic watchmaker (Marshall), believing that his wife is cheating on him (she isn't), rigs up a bomb set to go off in the cellar at 4:00 P.M. Ransacking teenagers trap Paul and tie him in the basement next to his own bomb . . . which doesn't go off at the appointed hour, since the electric current in the house is off!

9. *The Perfect Crime,* AHP. CBS, 20 Oct. '57. Script by Stirling Siliphant from Ben Ray Redman's story. Vincent Price, James Gregory. A variation on the exchange of guilt theme: a criminologist (Price) finds out that he had helped send an innocent man to jail.

10. *Lamb to the Slaughter,* AHP. CBS, 13 Apr. '58. Script by Roald Dahl from his own short story of the same title. Barbara Bel Geddes, Allan Lane, Harold J. Stone. The most famous of all Hitch TV films. In anger, Mrs. Maloney (Geddes) kills her police-chief husband by bludgeoning him with a leg of mutton. The police investigate, but can find no blunt instrument around his premises because Mrs. Maloney, *and* the cops, devour the leg of mutton while they're snooping around.

11. *Dip in the Pool,* AHP. CBS, 14 Sept. '58. Script by Francis Cockrell from Roald Dahl's short story of the same title. Keenan Wynn, Louise Platt, Philip Bournef, Fay Wray. Cheap chiseller Rotibol (Wynn) likes to bet everyone on how far their ship will travel each given day; when he attempts to cheat his shipmates he becomes a victim of his own cunning.

12. *Poison,* AHP. CBS, various dates in late '58. Script by Casey Robinson from Roald Dahl's short story. James Donald, Wendell Corey, Arnold Moss, Weaver Levy. A clever indictment of psychoanalysis; Harry, an alcoholic (Donald), tells the doctor and everyone else who will listen that he's being attacked by a deadly snake.

13. *Banquo's Chair,* AHP. CBS, 3 May '59. Script by Francis Cockrell from Rupert Croft-Cooke's story. John Williams, Kenneth

Haigh, Max Adrian, Reginald Gardiner. My personal favorite among Hitch's TV films. Inspector Brent (Williams), trying to solve the murder of a spinster, hires an actress to play the spinster's "ghost," confront the suspects and flush out the murderer. After the frightened murderer confesses, Brent discovers that the actress was unable to arrive; what had confronted the suspects was the spinster's *real* ghost!

14. *Arthur*, AHP. CBS, 27 Sept. '59. Script by James P. Cavanaugh from Arthur Williams' story. Laurence Harvey, Hazel Court, Robert Douglas, Barry Harvey. Actually a switch on *Lamb to the Slaughter* (above), in that both deal with a murderer's suppressing the evidence successfully. Helen (Court) annoys Arthur (Harvey), who doesn't want to get married, so Arthur grinds her up and feeds her to his chickens. Hitchcock's protege, director Robert Douglas, played a police inspector.

15. *The Crystal Trench*, AHP, CBS, 4 Oct. '59. Script by Stirling Silliphant from A. E. W. Mason's short story. James Donald, Patricia Owens. Yet another variation on the director's obsession with stories about the interchangability of guilt and innocence: young Mrs. Ballister (Owens) *appears* to feel deeply about the loss of her husband, who was killed in a crevasse while mountain-climbing.

16. *Incident at a Corner*, FST. NBC, 5 Apr. '60. Script by Charlotte Armstrong from her novelette of the same title. Vera Miles, Paul Hartman, George Peppard. With *Revenge* (above) and the feature *The Wrong Man*, this is the third film directed by Hitch in which Vera Miles plays a hallucinatory (or *perhaps* hallucinatory) woman. When a girl (Miles) tells everyone she's spotted her dead brother-in-law at a bus stop, no one believes her. Her sister remarries despite evidence hubby's still alive, and at the end the girl sees the dead man *again*. The plot remains unresolved (as does so much of Hitch's work).

17. *Mrs. Bixby and the Colonel's Coat*, AHP, NBC, 27 Sept. '60. Script by Halsted Welles from Roald Dahl's short story. Audrey Meadows, Les Tremayne, Stephen Chase, Sally Hughes, Bernie Hamilton, Lillian Culver, Harry Cheshire. Mrs. Bixby (Meadows), victim of a humdrum marriage, tells her hubby (Tremayne) that she's visiting her aunt overnight . . . the plot's "weenie" deals with the evidence of her subsequent infidelity.

18. *The Horseplayer*, AHP. NBC, 14 Mar. '61. Original script by Henry Slesar. Claude Rains, Ed Gardner. The plot is simple: grateful horseplayer Sheridan (Gardner) donates most of his winnings to the Church, which interests Father Amion (Rains). The French lunatic fringe would probably make much of Hitchcock's attitude toward Catholicism as revealed in this one.

19. *Bang! You're Dead!*, AHP. NBC, 17 Oct. '61. Script by Harold

Swanton from Margery Vosper's story. Biff Elliott, Lucy Prentiss, Billy Mumy, Steve Dunne. Unbeknownst to his family, that gun which little Jackie (Mumy) is toting around the house is a *real* one.

20. *I Saw the Whole Thing,* AHH. CBS, 11 Oct. '62. Script by Henry Cecil and Henry Slesar from Slesar's original. John Forsythe, Kent Smith, Evan Evans, Philip Ober, John Fielder, Claire Griswold. A very complex and quiet plot based entirely on the assignment of guilt. Mystery writer Barnes (Forsythe) proves that the witnesses are unreliable when he defends himself in court after being charged with hit-and-run . . . His wife, not Barnes, himself, proves to be guilty of the offense.

Analysis of the Plane and Cornfield Chase Sequence in NORTH BY NORTHWEST

What does a moviegoer think of when he's asked to name some moments of pure Hitchcockian suspense or terror? More than likely he will recall some outrageous, frightening, highly visual, almost wordless sequence like the shower murder in *Psycho,* or the chase around the presidential faces in *North by Northwest,* or, in the same film, the pursuit of Cary Grant across a lonely midwestern cornfield by an armed plane. This last has been widely regarded as the archetypal Hitchcockian suspense situation and the summation of his many chase sequences. An innocent figure is isolated in a banal setting that suddenly explodes with violence, terror, and death; almost all the possibilities of self-defense or running for cover are exhausted before the chase even begins.

Hitchcock's skill lies chiefly in the art of editing, of controlling the visual rhythms and tempo of his films, to involve the viewer subjectively in the experience. The sketches and descriptions which follow cannot recreate the immediate impact of the sequence, but they can show how such an impact was created and why it is so powerful. To explain his art Hitchcock has regularly appealed to basic principles of montage, particularly montage "to create ideas" or "to create violence and emotion"—both of which are in full evidence here. To insure audience identification, he stresses his theory of the "subjective" side of montage, "the juxtaposition of imagery relating to the mind of the individual," in this case Grant and the audience as his surrogate.

You have a man look, you show what he sees, you go back to the man. You can make him react in various ways. You see, you can make him look at one thing, look at another—without his speaking, you can show his mind at work, comparing things—any way you run there's complete freedom. It's limitless, I would say, the power of cutting and the assembly of the images.[1]

Hitchcock could be speaking in particular about the long opening wait after Grant's arrival by bus. It is here that Hitchcock "hooks" us—through the skillful alternation of subjective and objective shots, their timing and flow, the rhythms of action and reaction, the ominous silence, Grant's apprehension and bewildered expressions, the gradual canceling out of all obvious possibilities —all of which opens the way for our acceptance of the outrageous airplane attack and the explosive finale.

Two other important aspects of Hitchcock's artistry are revealed here. Within each shot there is a formal stylized note; when stopping the projector to draw each shot, there was usually a "perfect" moment when it should be done, when the harmonies and oppositions of horizontal and vertical lines and of angles seemed just right. Hitchcock never forgets the two-dimensional quality of film and the nature of the frame: "You have a rectangle to fill. Fill it. Compose it." No dead space. Hitchcock's stylizing quality, which probably dates from his earliest years in film when he designed titles and drawings on titles for silent films, is never sacrificed to the temporal rhythms of his "action" editing, but nevertheless leaves an impression of artistic detachment and control that mingles our terror with aesthetic distance and satisfaction.

Hitchcock is also fond of citing this sequence as a prime example of "avoiding the cliché." Instead of sending Grant to a dark alley to be done in, he sends him to a dull prairie bus stop in broad daylight. (See p. 22 of the interview in part one of this volume.) The resulting terror is all the more startling; the drabness of the prairie suddenly becomes a dangerous openness; man is turned into an animal with no place to run for cover. The ordinary becomes mysterious and potentially terrifying. The attack of the airplane is, of course, outrageous and carries with it a certain surreal tone, evidence of a universe gone berserk. But at the same time, such exaggerated images, avoiding the cliché, restore us to the delights of elemental chase scenes which were the essence of primitive films. In both subject matter and method, then, this sequence is a kind of "pure" cinema.

The sequence also profits from its context in the narrative line of the film thus far. Grant has been told by Eva Marie Saint, who has appeared to be his ally but in reality is the mistress of the man who wants him killed, that he is to meet a certain Kaplan at this lonely

[1] All quotations are from the introduction to Peter Bogdanovich, *The Cinema of Alfred Hitchcock* (New York: The Museum of Modern Art, 1962).

bus stop. Grant has been mistaken for Kaplan, kidnapped as Kaplan, involved in a murder as Kaplan, and now hunted both as Kaplan and his real self. Now he thinks the real Kaplan will appear and clear up the mystery, enabling him to clear himself. We share his quizzical and nervous attitude about what will happen, but smile at his naïveté, knowing he is in a more dangerous situation than he realizes. We know Kaplan does not exist. We have already seen Eva Marie Saint help Grant out—by telephoning the evil "Mr. Leonard" (Martin Landau) a few phone booths away and receive instructions to send him to his death. To increase our suspense, Hitchcock hasn't allowed us to hear the conversation and instructions; instead he has underlined Saint and Landau's complicity and her betrayal of Grant by panning in long shot from the booth where she dials the call to the booth where he receives it. Hence we share Grant's anxiety, but in a more complicated, dizzying way. (An additional complicating factor is that somehow we *do* believe that she is really in love with Grant and can't be following Landau's instructions.)

Within the sequence there is also a great variety of shading and subtlety which we hope the drawings will reveal. It is not just a long wait until the violence of the plane bursts upon us. Instead Hitchcock gives us a series of build-ups and false starts which he usually dissolves in comedy; throughout the sequence, his own subtlety is superbly complemented by Grant's excellent under-playing of the whole scene. There is the first drama: the long, silent opening section, the gradual appearance of cars (which might contain Kaplan), the end of such expectations in the comedy of Grant wiping the dust from a passing car from his eyes. This is followed by another build-up, the appearance of the other man from the field across the road, the sizing up of this unlikely prospect, the synchronous double-tracking shots in which Grant approaches him, his comic farmer's laconic responses which disqualify him as Kaplan and his obvious self-possession which contrasts with Grant's growing nervousness. As he disappears on the bus, his chance remark prepares for the major attack of the plane on Grant. "That's funny, that plane's dustin' crops where there ain't no crops." Then the sequence bursts wide open in movement and in space as the plane pursues Grant. The famous reverse tracking shot of Grant running from the pursuing plane unhinges us from any stable feelings we might have had. The cutting then tends to get quicker and quicker—especially in the moments before Grant is nearly destroyed and then saved by the truck he halts.

Even amid the terrors of the attack, Hitchcock never quite abandons the comic note. Hiding in the corn, Grant manages a smile thinking he has outwitted his pursuers and for a moment he even seems to enjoy the game; but they return with poison gas, forcing him out of his hiding place. After the plane hits the truck, the men escaping from the exploding vehicle run like Mack Sennett figures into the corn. And Hitchcock ends the sequence with a comic view of a bow-legged farmer futilely chasing his pick-up truck which Grant has stolen—the truck disappearing on the infinite road which we first saw in this way at the opening.

While the end fades into comedy, the basic note of terror and apprehension is nonetheless maintained and heightened. Bernard Herrman's superb music, which begins with the explosion of the plane and truck, marks a certain finality to the episode. We now know Grant is out of danger and we are relieved; the music seems to underline the events of the explosion, the cumulative effect of the episode, its rhythmic conclusion. But it also has a nervous jagged tension which lifts the images to an apocalyptic level that promises worse things to come. Hitchcock allows us to withdraw from the terror and to become onlookers by a series of long-held stylized shots in which little action—other than the burning truck and plane—occurs (we see it through the farmer's arm, then we see the onlooker's backs as they watch it). Grant's own movements are slow and stylized and characterized by a return to deliberation and control. His escape with the truck exploding behind him has a slight effect of slow motion, perhaps because the music dominates any sense of very individual action or explosive sounds. A few moments later he quietly backs to the position in the rear of the spectators as they move forward; there is a feeling of balance in the movement. Once again he is in control but propelling the movie into new action as he steals the pick-up truck and zooms off. Despite the relief we feel, the images of exploding plane and truck, of an apocalyptic finale, are central and enduringly present.

In the drawings that follow, much of the effect of the editing and visual rhythms can be approximated by simply glancing at them. Figures in italic at the end of the notations indicate the approximate length in seconds of each shot. A small "a" or "b" with a shot means that it is the same shot but that enough new action was taking place in it to warrant two drawings. This was most frequently true, of course, of tracks and pans, but it occasionally happened in shots in which the camera did not move. It should also be mentioned that unfortunately we worked with a badly spliced print and that there was a reel break between shots 34 and 35, with most of the splices occurring near the end and opening of the reels. Consequently we may be missing something and the times of shots may not be exact.

Finally, the artists and editor would like to add a personal note: We enjoyed this work immensely and hope it proves of value.

Through it we grew to have a great respect for Hitchcock and to understand the bases of his craft. Hitchcock is a supremely visual artist whose work needs no apology—not even the abstract glorifications of *Cahiers'* critics. Lastly, we became aware of the perfect suiting of acting, timing, delivery of dialogue, gesture, etc., of Cary Grant to the rhythms of Hitchcock's editing.

SEQUENCE AND ANALYSIS OF SHOTS

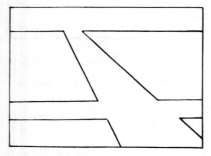

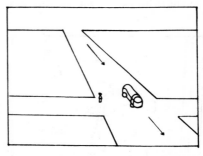

1. Extreme long shot, aerial. Dissolve from Chicago train station, shot of Eva Marie Saint's face, as she tells Grant to hurry from nonexistent police, to empty road across fields where rendezvous with Kaplan is to take place. We see and hear bus arriving, door opening.

1b. Grant emerges, bus leaves, Grant is alone. *52 sec.*

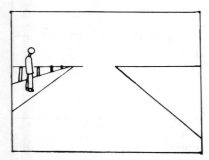

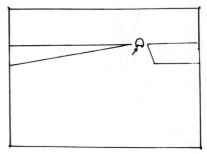

2. Long shot, low angle. Grant at roadside, waiting. *5*

3. Long shot. Main road, looking down it from Grant's eyes, bus going away in distance. *4*

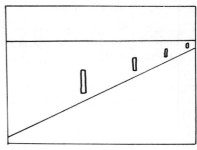

4. Medium shot. Grant by sign, turns left to right, looking for someone. *3⅔*

5. Long shot. View across road, empty fields with posts, seen through Grant's eyes. *4*

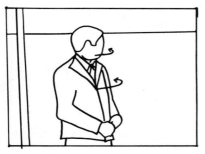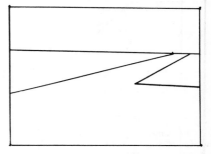

8. Medium shot. Grant by sign again, waiting, turns head and looks behind him. *3*

9. Long shot. The field behind him, road, Grant's view. *4*

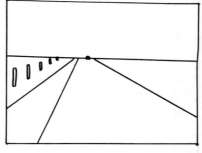

12. Medium shot. Grant by sign, turns right. *2*

13. Long shot. Empty main road, car coming in distance. *4*

6. Medium shot. Grant by sign, turns from right to left, looking. *3*

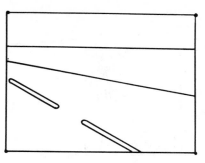

7. Long shot. Across fields, Grant's view. *4*

10. Medium shot. Grant by sign, waiting, turning forward—long waiting feeling. *6½*

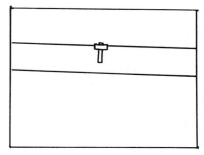

11. Long shot. Empty landscape across road, signs, posts. *3*

14. Medium shot. Grant by sign, looking at car approaching. *¾*

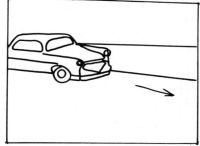

15. Medium long shot. Car goes by fast, whizzing sound, camera pans slightly to left. *1¾*

16. Medium shot. Grant by sign, moves back to left, follows car with his eyes. *2*

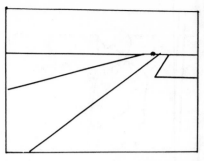

17. Long shot. Road seen through his eyes, car going, sound recedes. *4*

20. Medium shot. Grant by sign, same pose, still looking, turns from right to left. *2¾*

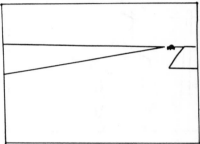

21. Long shot. Road, car in distance, sounds begin. *3½*

24. Medium shot. Grant by sign, looking at car coming. *3*

25. Medium long shot. Car closer, rushes past, camera pans a bit to follow it. *3*

18. Medium shot. Grant by sign, hands in pockets, turns from left to right. *3*

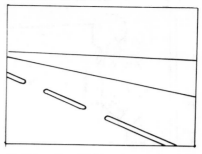

19. Long shot. Field across road again, Grant's view. *3*

22. Medium shot. Grant by sign, looking at car, no movement. *2½*

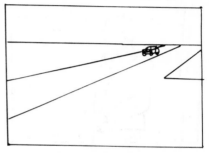

23. Long shot. Car coming closer, sound increasing. *3¾*

26. Medium shot. Grant by sign, takes hands from pockets, turns left to right to follow car. *3*

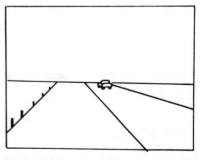

27. Long shot. View of road, car in distance receding. *3¾*

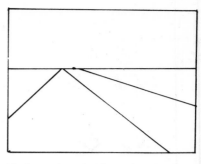

28. Medium shot. Grant by sign, waiting again. 2¼

29. Long shot. Road, truck coming; we hear its sound. *4*

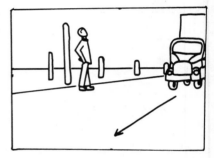

32. Medium shot. Grant by sign. 2¼

33. Medium long shot. Truck whizzes by.

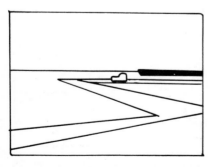

35. Very long shot. Fields across way, car coming out behind corn. *5?*

36. Medium shot. Grant by sign puzzled by car. *5*

30. Medium shot. Grant by sign, sound of truck increasing. *3*

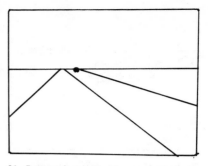

31. Long shot. Truck coming down road, sound still increasing. *3¾*

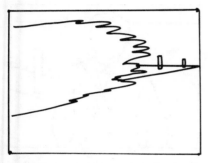

33b. Kicking up dust, obscuring Grant, camera pans slightly left, and Grant emerges out of the dust gradually. *4*

34. Medium shot. Grant wiping dust from his eyes, turns to right. *7?*

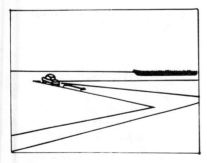

37. Long shot. Car making turn on dirt road. *4*

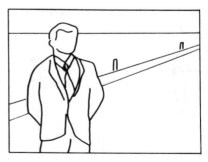

38. Medium shot. Grant awaiting car. *3⅔*

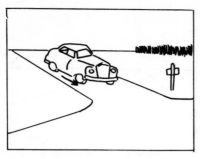

39. Long shot. Car nearing main road, camera pans following it to right, a sign there. *4*

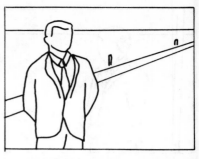

40. Medium shot. Grant waiting to see what will happen. *3⅓*

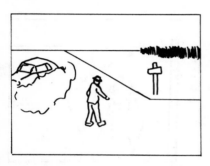

43. Long shot. Sound of car turning around, dust raised, car turns around and the man walks towards main road opposite Grant, looking back at the car leaving. *1⅘*

44. Medium shot. Grant, closer than previous shots, eyeing the man. *1½*

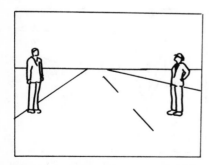

47. Long shot, low angle. Road in the middle stretching to infinity, two men oddly stationed on either side of road, the other man looking up the road a bit. *7*

48. Medium shot. Grant's reaction, takes hands out of pockets, opens coat, puts hands on hips, contemplates situation. *6⅓*

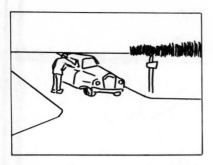

41. Long shot. Man getting out of car, talking to driver, we hear the door of car slam. *3½*

42. Medium shot. Grant reacting, wondering, getting ready to meet this man.

2

45. Long shot. Camera pans right slightly; the man goes over by the sign and turns his head to look up the road and over at Grant. *4⅓*

46. Medium shot. Same as 44. Grant's reaction, his head tilts and he looks across the road. *3⅘*

49. Long shot. Same as 45, only the man turns his head looking the other way. *3½*

50. Medium shot. Same as 48, but Grant has both hands on hips now, his head looking across. His head turns up road to see if anyone is coming; he looks back, one hand on hip, other at side, at man across way.

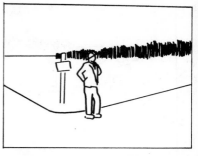

50b. Starts walking across road part way. *10*

51. Long shot. Same as 49. The man on the other side of road, but seen from Grant's viewpoint as he crosses the road; camera tracks across road part way, acting as his eyes. *2⅖*

54. Medium shot. Grant on other side of road but camera tracks to continue movement of 50, 52.

54b. Camera continues tracking to other side of road until other man comes into view and Grant nervously begins to talk to him. Grant's hands are a bit nervous in movement; he plays with his little finger; the other man has hands in pockets.

> *Grant:* (after a long wait) Hi! (a long pause follows) Hot day. (Another pause.)
> *Man:* Seen worse.
> *Grant:* (after a long pause) Are you supposed to be meeting someone here?
> *Man:* Waiting for the bus, due any minute.
> *Grant:* Oh! (another pause.)
> *Man:* Some of them crop duster pilots get rich if they live long enough.
> *Grant:* Yeah! (very softly). *21*

52. Medium shot. Grant, seen objectively, walks across road; synchronous tracking camera continuing movement begun in 50b.

53. Medium long shot. Same as 49, 51 of other man across road, but camera tracks in on him, acting as Grant's eyes, continues movement begun in 50b.

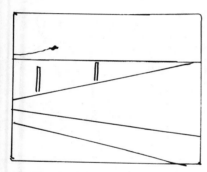

55. Long shot. Fields with plane at great distance in far left of frame coming right. 2⅔

56. Medium shot. Reaction shot of both looking at plane.
    Grant: Then . . . a . . . (pause) then your name isn't Kaplan?
    Man: Can't say that it is cause it ain't. (Pause) Here she comes (as he looks down the road). *11*

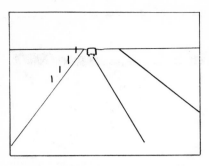

57. Long shot. Bus coming down the road.

> *Man:* (voice off) . . . right on time.
> 2⅔

58. Medium. Same as 56, the two of them talking, then looking again across road at crop duster.

> *Man:* That's funny.
> *Grant:* (very softly) What?
> *Man:* That plane's dusting crops where there ain't no crops.
> Grant turns to look. *8*

61. Medium. Bus arriving and coming quite close to camera. *1⅘*

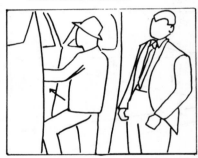

62. Medium. Man gets on as door of bus opens and seems to shut Grant out. The bus leaves.

64. Medium. Grant in front of road by sign, puzzled and rather innocent looking; sound of plane approaching. *2⅓*

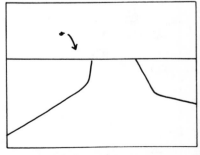

65. Long. Plane coming toward camera, which acts as Grant's eyes, still far but closer and with sound increasing. *3⅘*

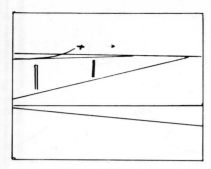

59. Long shot. Same as 55, field with the plane over it as they see it. *4*

60. Medium. Two men off center looking at plane; Grant's hands continue nervous movements; the others are in his pockets as before. **Sound of approaching bus.** *3⅓*

62b. Grant puts hands on hips and (a) looks across, then (b) at watch. For a second he is alone in the frame as the bus goes out of sight. *23⅓*

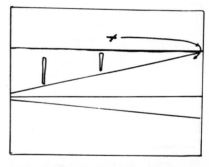

63. Long shot. Same as 59, what Grant sees across road; the plane goes to end of frame and turns right, toward him. *5⅕*

66. Medium. Same as 64, Grant reacting. *2¼*

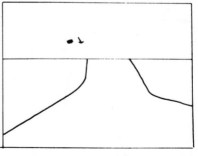

67. Long. Same as 65 but plane closer and louder. *2⅛*

68. Medium close. Closer shot of Grant, still puzzled and confused as plane comes at him. *4½*

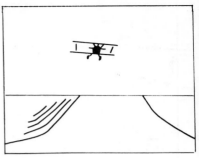

69. Long. Plane clearly coming at him, filling mid-frame, very loud. *1⅓*

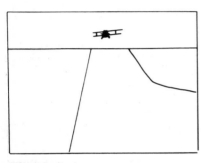

72. Long. Plane going away from him. *3*

73. Medium. Grant on ground getting up, kneeling on left knee. *3½*

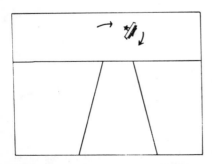

76. Long. Plane in distance banking. *2⅓*

77. Medium. Grant up and about to run. *2⅓*

70. Medium. Grant drops, a short held shot, he falls out of frame at bottom. ⅔

71. Medium. Grant falling on ground, both arms on ground, plane behind him, he in a hole. 3½

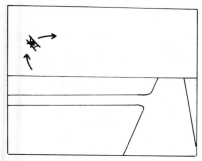

74. Long. Plane going farther away and sound receding. 2⅘

75. Medium. Grant getting up. 2⅕

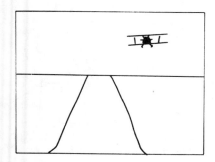

78. Long. Plane approaching again, sound getting louder. 2

79. Medium. Grant runs and falls in ditch. 1½

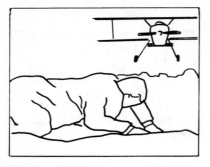

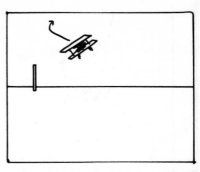

80. Medium. Grant in ditch, sound of plane and bullets sprayed on him, smoke; he turns head to left and faces camera to watch when plane is gone. *5⅓*

81. Long. Plane getting ready again, banking. *5⅕*

84. Medium. Same as 82, Grant rising from ditch, receding plane sound. *1½*

85. Long, low angle. Grant runs to road to try to stop car.

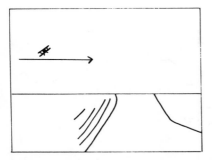

87. Long. Plane in distance, sounds again. *2⅓*

88. Medium long. Grant's back with plane in distance at far left coming at him.

82. Medium. Grant in ditch coming up, gets up on left arm, sound of receding plane. 2½

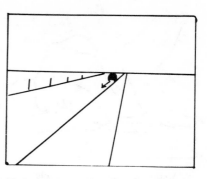

83. Long. The road as Grant sees it, car in distance. 2½

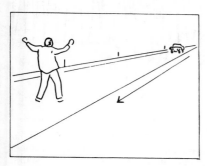

85b. He tries to flag it down. Car sounds approach and it whizzes by. 9½

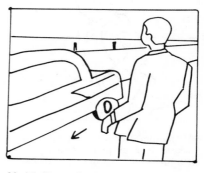

86. Medium. Grant's back after turning left as the car whizzes by. 4⅔

88b. Medium. He looks at plane, turns around looking for a place to hide, looks at plane again, turns around and runs towards camera. Camera reverse tracks.

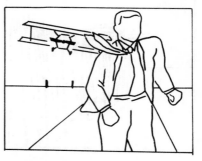

88c. Grant running toward camera, camera reverse tracking. He turns around twice while running to look at plane; it goes over his head just missing him. 13½

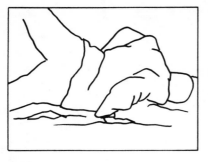

89. Medium. Grant falling, side view, legs up, bullet and plane sounds. *5*

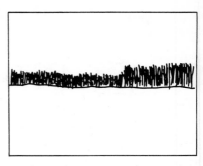

90. Long. Cornfield as Grant sees it, a place to hide. *2½*

93. Medium. Grant running, turns back to look at plane, camera tracks with him as he runs to cornfield. *4½*

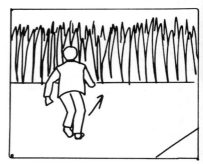

94. Long, low angle. Grant's back as he runs into cornfield; low camera angle shows lots of ground, stalks. Grant disappears into the corn. *2¾*

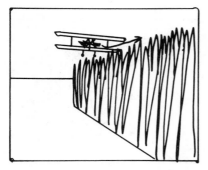

97. Long. Plane coming along edge of cornfield and over it; it gets very loud. *4½*

98. Medium. Same as 96. Corn rustles, wind from the plane blows over. Grant sees he's out of danger, and smiles a bit, feeling that he's outwitted his pursuers. *15⅘*

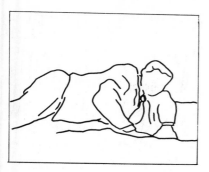

91. Medium. Grant lying flat on ground, looking. *1⅔*

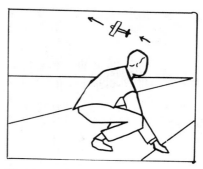

92. Medium. Grant getting up, plane in distance going around corner, banking again for new attack. *3*

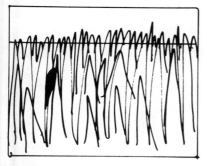

95. Medium. Picture of corn; a patch reveals where Grant is hiding. The corn rustles. *2*

96. Medium. Camera follows Grant down as he falls on ground inside the corn patch. He turns back to look up to see if plane is coming. A cornstalk falls; then he looks down again, up again, down again, up. *7¾*

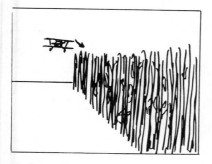

99. Long. Plane coming in on bend, repeating pattern of 97; it gets louder. *3¼*

100. Medium. Same as 96, 98. Grant in corn, getting up, looking around, suddenly aware of plane in new way; he's startled that it's coming back. *4⅘*

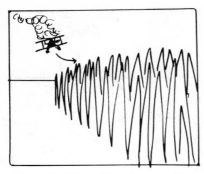

101. Long. Plane over corn, repeating pattern of 97, only smoke coming out of it; plane comes closer to camera. *7*

102. Medium. Same as 96, 98, 100; Grant's reaction to smoke, smoke fills up screen; Grant coughs, takes out handkerchief, camera follows him as he raises himself up and down; coughing sounds. *12¼*

105. Medium. Grant in corn, same as 103, but he is standing all the way up; he moves forward, looks back up for plane, makes dash for the truck coming down road; he goes out of frame for a moment at the end. *3½*

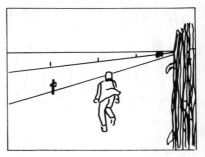

106. Long. Grant running toward truck, gets to road from corn; truck farther along the road; sounds of truck. *4*

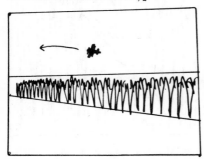

109. Long. Continuation of 107, plane further left. *2*

110. Medium. Grant trying to stop truck, sounds of horns, brakes; (appears to be back projection). *2*

103. Medium. Grant in corn, new shot; he runs towards camera trying to get out of corn; rustling corn, he looks out of field. *4½*

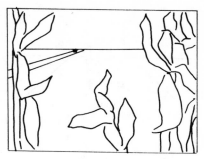

104. Long. View out of corn; Grant's view of the road as framed by the corn; tiny speck on road in distance is truck. *2¾*

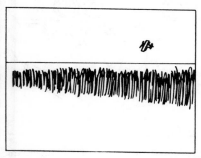

107. Long. Plane banking over corn, getting ready to turn toward him; faint plane sounds; horn of truck. *2*

108. Medium long, low angle. Grant in road, truck coming, he waves at it. *1⅔*

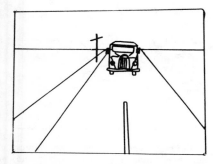

111. Long. Truck approaching, getting bigger. *2*

112. Medium. Same as 110, but he looks at plane coming in on his left, then puts up both hands instead of one, and bites his tongue. *1½*

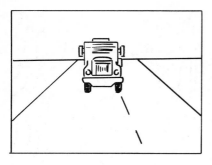

113. Long. Truck even closer, about 50 feet in front of camera. *1*

114. Medium. Same as 112, 110. Grant waving frantically now. *1+*

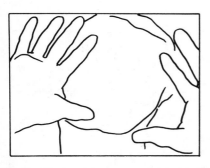

116b. His hands go up and his head goes down. *4/5*

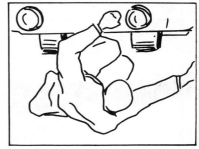

117. Medium. Grant falls under the truck, front view. *1—*

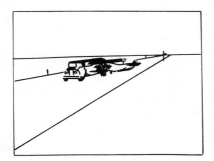

120. Long. Plane hits truck, view from across road. *1*

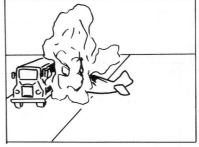

121. Long. Truck bursts into flames; another angle of truck and plane; music begins and continues to end of scene. *2½*

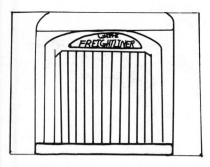

115. Close-up. Grille of truck as it tries to halt; brake sounds. *1—*

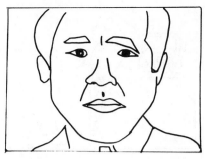

116. Close-up. Grant's face in anguish about to be hit.

118. Medium. Grant under the truck, side view. *2¼*

119. Long low. Plane comes toward truck (and camera). *1¾*

122. Long. Shot of truck in flames from in front, two men exit hurriedly from cab.

    One: Let's get out of here. It's going to explode. *6½*

123. Long. Backs of men running somewhat comically to cornfield. *2—*

124. Long. Grant runs toward camera from explosions of oil truck behind him. Music tends to mute explosion sounds. *2¾*

125. Long. Reverse angle of Grant as he now backs away from explosion. A car has just pulled to side of road, followed by a pickup truck with a refrigerator, which pulls to the side in front of it. Doors open and people get out. Grant goes over—now in very long shot—and talks to them. No sounds are heard (the music continues), but we see his motions of explanation. *9+*

128. Long. Backs of others watching explosion. *6⅓*

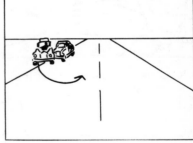

129. Very long. Grant takes the pickup truck with refrigerator in back and pulls out while they are watching the explosion. *3⅘*

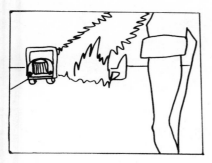

126. Long. View of explosion in distance with close-up of farmer's arm on right. 2—

127. Long. Grant and others watch explosion, Grant backing away from scene to right of frame while they all look forward; they all move closer to the wreck as Grant retreats unnoticed by them. 11

130. Long. Same as 128, but the farmer turns, seeing his car being taken, and shouts "Hey." 2½

131. Long. Chase by the farmer, bowlegged and comic; the farmer finally stops as the car recedes into the distance. "Come back, come back," he is shouting. Dissolve to truck on street in front of elegant Chicago hotel, the cops puzzling over it. 15

Filmography[1]

A more complete filmography with listings of casts, writers, editors, etc., can be found in Robin Wood's *Hitchcock's Films* and in Truffaut's *Hitchcock*.

EARLY WORK AS ASSISTANT DIRECTOR AND ART DIRECTOR

The following films were all directed by Graham Cutts.

Woman to Woman. 1923.
The White Shadow. 1923.
The Passionate Adventure. 1924.
The Blackguard. 1925.
The Prude's Fall. 1925.

As Director

Number 13. 1922 (unfinished).
Always Tell Your Wife. 1922 (part codirected with Seymour Hicks).
The Pleasure Garden. 1925. Sc: Eliot Stannard. S–novel: Oliver Sandys. Ph: Baron Ventimiglia. Pro: Gainsborough.
The Mountain Eagle. 1926. Sc: Eliot Stannard. Ph: Baron Ventimiglia. Pro: Gainsborough. Pl: Bernard Goetzke, Nita Naldi.
The Lodger. 1926. Sc: Alfred Hitchcock, Eliot Stannard. S–novel: Mrs. Belloc-Lowndes. Ph: Baron Ventimiglia. Pro: Gainsborough. Pl: Ivor Novello.
Downhill. 1927. Sc: Eliot Stannard. S–play: Ivor Novello and Constance Collier. Ph: Claude L. McDonnell. Pro: Gainsborough, 1927, G. B. Pl: Ivor Novello, Ben Webster, Isabel Jeans.

[1] Code: Sc—Screenplay
S—Source in a ———
Ph—Photography
Pro—Production
M—Music
Pl—Players

Easy Virtue. 1927. Sc: Eliot Stannard. S–play: Noel Coward. Ph: Claude L. McDonnell. Pro: Gainsborough Prod. Pl: Isabel Jeans, Franklin Dyall.

The Ring. 1927. Sc: Alfred Hitchcock. Ph: Jack Cox. Pro: British International Pictures, 1927, G. B. Pl: Carl Brisson, Lillian Hall-Davies, Ian Hunter.

The Farmer's Wife. 1928. Sc: Alfred Hitchcock. S–play: Eden Philpotts. Ph: Jack Cox. Pro: British International Pictures, 1928, G. B. Pl: Lillian Hall-Davies, James Thomas.

Champagne. 1928. Sc: Eliot Stannard. S–story: Walter C. Mycroft. Ph: Jack Cox. Pro: British International Pictures, 1928, G. B. Pl: Betty Balfour, Gordon Harker.

Harmony Heaven. 1929 (part directed). Pro: British International Pictures, 1929.

The Manxman. 1929. Sc: Eliot Stannard. S–novel: Sir Hall Caine. Ph: Jack Cox. Pro: British International Pictures, 1929, G. B. Pl: Carl Brisson, Anny Ondra, Malcolm Keen.

Blackmail. 1929. Sc: Alfred Hitchcock, Benn W. Levy, Charles Bennett. S–play: Charles Bennett. Ph: Jack Cox. Pro: British International Pictures, 1929. Pl: Anny Ondra (with the voice of Joan Barry), Sara Allgood, John Longden.

Elstree Calling. 1930. Part directed only. Supervising Director: Adrian Brunel. Sc: Val Valentine. Ph: Claude Friese-Greene. Pro: British International Pictures, 1930. Pl: Anna May Wong, Donald Calthrop, Gordon Harker.

Juno and the Paycock. 1930. Sc: Alfred Hitchcock, Alma Reville. S–play: Sean O'Casey. Ph: Jack Cox. Pro: British International Pictures, 1930. Pl: Sara Allgood, Edward Chapman, Marie O'Neill.

Murder. 1930. Sc: Alma Reville. S–play: *Enter Sir John* by Clemence Dane and Helen Simpson. Ph: Jack Cox. Pro: British International Pictures, 1930. Pl: Herbert Marshall, Norah Baring, German language version *Mary* directed by Hitchcock with Alfred Abel and Olga Tchekowa.

The Skin Game. 1931. Sc: Alfred Hitchcock and Alma Reville. S–play: John Galsworthy. Ph: Jack Cox. Pro: British International Pictures, 1931. Pl: Edmund Gwenn, Jill Esmond, John Longden.

Rich and Strange (USA—*East of Shanghai*). 1932. Sc: Alma Reville, Val Valentine. S–theme: Dale Collins. Ph: Jack Cox, Charles Martin. Pro: British International Pictures, 1932. Pl: Henry Kendall, Joan Barry.

Number Seventeen. 1932. Sc: Alfred Hitchcock. S–play and novel: Jefferson Farjeon. Ph: Jack Cox. Pro: British International Pictures, 1932. Pl: Leon M. Lion, Anne Grey.

Lord Camber's Ladies. 1932. Hitchcock acted only as producer; Director: Benn W. Levy.

Waltzes from Vienna (USA—*Strauss' Great Waltz*). 1933. Sc: Alma Reville, Guy Bolton. S–play: Guy Bolton. Pro: Gaumont British, by G. F. I., 1933, G. B. Pl: Jessie Matthews, Esmond Knight.

The Man Who Knew Too Much. 1934. Sc: A. R. Rawlinson, Edwin Greenwood. S–theme: Charles Bennett, D. B. Wyndham Lewis. Ph: Curt Courant. Pro: Gaumont British Pictures, Great Britain. Pl: Leslie Banks, Edna Best, Peter Lorre.

The Thirty-Nine Steps. 1935. Sc: Alma Reville, Charles Bennett. S–novel: John Buchan. Ph: Bernard Knowles. Pro: Gaumont British. Pl: Robert Donat, Madeleine Carroll.

The Secret Agent. 1936. Sc: Charles Bennett. S–play: Campbell Dixon and novel: *Ashenden* by Somerset Maugham. Ph: Bernard Knowles. Pro: Gaumont British, 1936. Pl: Madeleine Carroll, John Gielgud, Peter Lorre.

Sabotage. 1936 (USA—*A Woman Alone*). Sc: Charles Bennett. S–novel: *The Secret Agent* by Joseph Conrad. Ph: Bernard Knowles. Pro: Shepherd, Gaumont British Pictures, 1936. Pl: Sylvia Sidney, Oscar Homolka, John Loder.

Young and Innocent. 1937 (USA—*A Girl Was Young*). Sc: Charles Bennett, Alma Reville. S–novel: Josephine Tey. Ph: Bernard Knowles. Pro: Gainsborough, Gaumont British, 1937. Pl: Nova Pilbeam, Derrick de Marney, Percy Marmont.

The Lady Vanishes. 1938. Sc: Sidney Gilliat, Frank Launder. S–novel: *The Wheel Spins* by Ethel Lina White. Ph: Jack Cox. Pro: Gainsborough Pictures, 1938, G. B. Pl: Margaret Lockwood, Michael Redgrave, Paul Lukas.

Jamaica Inn. 1939. Sc: Sidney Gilliat, Joan Harrison. S–novel: Daphne du Maurier. Ph: Harry Stradling, Bernard Knowles. Pro: Mayflower Productions, 1939, G. B. Pl: Charles Laughton, Maureen O'Hara, Leslie Banks.

Rebecca. 1940. Sc: Robert E. Sherwood, Joan Harrison. S–novel: Daphne du Maurier. Ph: George Barnes. M: Franz Waxman. Pro: David O. Selznick, USA, 1940. Laurence Olivier, Joan Fontaine, Judith Anderson.

Foreign Correspondent. 1940. Sc: Charles Bennett, Joan Harrison. Ph: Rudolph Mate. M: Alfred Newman. Pro: Walter Wanger, United Artists, 1940. Pl: Joel McCrea, Laraine Day, Herbert Marshall.

Mr. and Mrs. Smith. 1941. Sc: Norman Krasna. S–story: Norman Krasna. Ph: Harry Stradling. M: Roy Webb. Pro: RKO, 1941. Pl: Carole Lombard, Robert Montgomery.

Suspicion. 1941. Sc: Samson Raphaelson, Joan Harrison, Alma Reville. S–novel: *Before the Fact* by Frances Iles. Ph: Harry Stradling. M: Franz Waxman. Pro: RKO, 1941. Pl: Cary Grant, Joan Fontaine, Nigel Bruce.

Saboteur. 1942. Sc: Peter Viertel, Joan Harrison, Dorothy Parker. S–idea: Alfred Hitchcock. Ph: Joseph Valentine. M: Charles Previn, Frank Skinner. Pro: Universal, 1942. Pl: Robert Cummings, Priscilla Lane, Otto Kruger.

Shadow of a Doubt. 1943. Sc: Thornton Wilder, Alma Reville, Sally Benson. S–story: George McDonnell. Ph: Joseph Valentine. M: Dmitri

Tiomkin. Pro: Universal, 1943. Pl: Joseph Cotten, Teresa Wright, Macdonald Carey.

Lifeboat. 1943. Sc: Jo Swerling. S–story: John Steinbeck. Ph. Glen MacWilliams. M: Hugo Friedhofer. Pro: Kenneth MacGowan/20th-Century-Fox. Pl: Tallulah Bankhead, William Bendix, Walter Slezak, John Hodiak.

Bon Voyage. 1944. Sc: J. O. C. Orton, Angus McPhail. S–idea; Arthur Calder-Marshall. Ph: Gunther Krampf. Pro: M. O. I., 1944. G. B., British Ministry of Information. Pl: John Blythe, the Moliere Players.

Adventure Malgache. 1944. Ph: Gunther Krampf, Pro: M. O. P., 1944, G. B., British Ministry of Information. Pl: Moliere Players.

Spellbound. 1945. Sc: Ben Hecht. S–novel: *The House of Dr. Edwardes* by Francis Beeding, adapted by Angus McPhail. Ph: George Barnes. Dream sequences by Salvador Dali. M: Miklos Rozsa. Pro: Selznick International, 1945. Pl: Ingrid Bergman, Gregory Peck, Leo G. Carroll.

Notorious. 1946. Sc: Ben Hecht. S–theme: Alfred Hitchcock. Ph: Ted Tetzlaff. M: Roy Webb. Pro: Alfred Hitchcock/RKO, 1946. Pl: Ingrid Bergman, Cary Grant, Claude Rains.

The Paradine Case. 1947. Sc: David O. Selznick. S–novel: Robert Hichens, adapted by Alma Reville. Ph: Lee Garmes. M: Franz Waxman. Pro: Selznick International, 1947. Pl: Gregory Peck, Ann Todd, Charles Laughton, Ethel Barrymore.

Rope. 1948. Sc: Arthur Laurents. S–play: Patrick Hamilton, adapted by Hume Cronyn. Ph: (Technicolor) Joseph Valentine, William V. Skall. M: Leon F. Forbstein. Pro: Sidney Bernstein/Transatlantic Pictures/Warner Brothers, 1948. Pl: James Stewart, Farley Granger, John Dall.

Under Capricorn. 1949. Sc: James Bridie. S–novel: Helen Simpson, adapted by Hume Cronyn. Ph: (Technicolor) Jack Cardiff, Paul Beeson, Ian Craig, David McNeilly, Jack Haste. M: Richard Adinsell. Pro: Sidney Bernstein/Transatlantic Pictures/Warner Brothers, 1949. Pl: Ingrid Bergman, Joseph Cotten, Michael Wilding.

Stage Fright. 1951. Sc: Whitfield Cook. S–two stories: Selwyn Sepson, adapted by Alma Reville. Ph: Wilkie Cooper. M. Leighton Lucas. Pro: Alfred Hitchcock, Warner Brothers, 1950, G. B. Pl: Marlene Dietrich, Jane Wyman, Michael Wilding.

Strangers on a Train. 1951. Sc: Raymond Chandler, Czenzi Ormonde. S–novel: Patricia Highsmith, adapted by Whitfield Cook. Ph: Robert Burks. M: Dmitri Tiomkin. Pro: Alfred Hitchcock, Warner Brothers, U.S.A. Pl: Farley Granger, Ruth Roman, Robert Walker.

I Confess. 1952. Sc: George Tabori, William Archibald. S–play: Paul Anthelme. Ph: Robert Burks. M: Dmitri Tiomkin. Pro: Alfred Hitchcock, Warner Brothers, 1952. Pl: Montgomery Clift, Anne Baxter, Karl Malden.

Dial M for Murder. 1954. Sc: Frederick Knott. S–play: Frederick Knott. Ph: (Warner-color, 3-D) Robert Burks. M: Dmitri Tiomkin. Pro: Alfred Hitchcock, Warner Brothers, 1954. Pl: Ray Milland, Grace Kelly, Robert Cummings, John Williams.

Rear Window. 1954. Sc: John Michael Hayes. S–novelette: Cornell Woolrich. Ph: (Technicolor) Robert Burks. M: Franz Waxman. Pro: Alfred Hitchcock, Paramount, 1954. Pl: James Stewart, Grace Kelly, Wendell Corey, Thelma Ritter.

To Catch a Thief. 1955. Sc: John Michael Hayes. S–novel: David Dodge. Ph: (Technicolor) Robert Burks. M: Lyn Murray. Pro: Alfred Hitchcock, Paramount, 1955. Pl: Cary Grant, Grace Kelly.

The Man Who Knew Too Much. 1955. Sc: John Michael Hayes, Angus McPhail. S–story: Charles Bennett, D. B. Wyndham Lewis. Ph: (Technicolor) Robert Burks. M: Bernard Herrmann. Pro: Alfred Hitchcock/ Paramount, Filmwire Prod., 1955. Pl: James Stewart, Doris Day.

The Trouble with Harry. 1956. Sc: John Michael Hayes. S–novel: Jack Trevor Story. Ph: (Technicolor) Robert Burks. M: Bernard Herrmann. Pro: Alfred Hitchcock, Paramount, 1956. Pl: Edmund Gwenn, John Forsythe, Shirley MacLaine.

The Wrong Man. 1957. Sc: Maxwell Anderson, Angus McPhail. S–story: Maxwell Anderson. Ph: Robert Burks. M. Bernard Herrmann. Pro: Alfred Hitchcock, Warner Brothers, 1957. Pl: Henry Fonda, Vera Miles.

Vertigo. 1958. Sc: Alec Coppel, Samuel Taylor. S–novel: *D'Entre les Morts* by Boileau and Narcejac. Ph: (Technicolor) Robert Burks. M: Bernard Herrmann. Pro: Alfred Hitchcock, Paramount, 1958. Pl: James Stewart, Kim Novak, Barbara Bel Geddes.

North by Northwest. 1959. Sc: Ernest Lehman. Ph: (Technicolor) Robert Burks. M: Bernard Herrmann. Pro: Alfred Hitchcock, MGM, 1959. Pl: Cary Grant, Eva Marie Saint, James Mason.

Psycho. 1960. Sc: Joseph Stefano. S–novel: Robert Bloch. Ph: John L. Russell. M: Bernard Herrmann. Pro: Alfred Hitchcock/Paramount, 1960. Pl: Janet Leigh, Anthony Perkins, Vera Miles.

The Birds. 1963. Sc: Evan Hunter. S–story: Daphne du Maurier. Ph: (Technicolor) Robert Burks. Sound Consultant: Bernard Herrmann. Pro: Alfred Hitchcock/Universal, 1963. Pl: Rod Taylor, Tippi Hedren, Jessica Tandy.

Marnie. 1964. Sc: Jay Presson Allen. S–novel: Winston Graham. Ph: (Technicolor) Robert Burks. M: Bernard Herrmann. Pro: Alfred Hitchcock/Universal, 1964. Pl: Tippi Hedren, Sean Connery.

Torn Curtain. 1966. Sc: Brian Moore. Ph: (Technicolor) John F. Warren. M: John Addison. Pro: Alfred Hitchcock/Universal, 1966. Pl: Paul Newman, Julie Andrews.

Topaz. 1969. Sc: Brian Moore. S–novel: Leon Uris. Ph: (Technicolor) Jack Hildyard. Pro: Alfred Hitchcock/Universal, 1969. Pl: Frederick Stafford, John Forsythe, Michel Piccoli.

Frenzy. 1972. Sc: Anthony Shaffer. S–novel: *Goodbye Piccadilly, Hello Leicester Square* by Arthur Labern. Ph: Gil Taylor. M: Ron Goodwin. Pro: Alfred Hitchcock/Universal, 1972. Pl: Jon Finch, Alec McCowen, Barry Foster, Billie Whitelaw, Vivien Merchant, Anna Massey.

Selected Bibliography[1]

I. INTERVIEWS AND ACCOUNTS

Hitchcock (no interviewer listed)

"Director Hitchcock Tells Young Film Directors How Easy It Is." *Making Films in New York,* August 1968.

"Director's Problems." *Listener,* 1938.

"Hitchcock and the Dying Art: His Recorded Comments." *Film,* Summer 1966.

"Hitchcock Talks About Lights, Camera, Action." *American Cinematographer,* May 1967.

"Murder—with English on It." *New York Times Magazine,* March 3, 1957, pp. 17, 42. Interesting example of a kind of article Hitchcock did regularly for the *New York Times Magazine* and *This Week Magazine.*

"Pourquoi j'ai peur la nuit." *Arts,* no. 77 (June 1, 1960).

"A Talk with Alfred Hitchcock." *Action* 3, no. 3 (May–June 1968): 8–10.

"Why You Need Thrills and Chills." *This Week Magazine,* September 22, 1957.

By Interviewer

Bitsch, Charles, and Truffaut, François. "Rencontre avec Alfred Hitchcock." *Cahiers du Cinéma,* no. 62 (August–September 1956): 1–5.

Cameron, Ian, and Perkins, V. F. *Movie,* no. 6 (January 1963): 4–6. Reprinted in Andrew Sarris, ed., *Interviews with Film Directors.* Indianapolis: Bobbs–Merrill, 1967, pp. 199–207. Also available as an Avon paperback.

Havemann, Ernest. "We Present Alfred Hitchcock." *Theatre Arts* 50, no. 9 (September 1956): 27–28, 91–92.

Higham, Charles, and Greenberg, Joel. "Alfred Hitchcock," in *The Celluloid Muse: Hollywood Directors Speak.* Chicago: Regnery; London: Angus and Robertson, Ltd., 1971, pp. 86–103. A long talk by Hitchcock, repeating familiar stories and theories, but centering on the American

---

[1] This bibliography does not include the many collections of mystery stories Hitchcock has edited and for which he usually contributes a brief introduction, e.g., *Stories They Wouldn't Let Me Do on TV, Stories Not for the Nervous, Fear and Trembling, My Favorites in Suspense,* etc.

films and offering new material on his relations with producers, writers, and actors.

Hitchcock, Alfred. "Hitchcock: Gooseflesh Is His Aim," as told to Atra Baly. *New York Journal American,* September 23, 1959.

———. "Preface." *Cahiers du Cinéma* no. 39 (October 1954): 11–13. Special Hitchcock issue.

Jameson, Barbara Birch. "3-D Spells Murder for Alfred Hitchcock." *New York Times,* October 11, 1953. On the problems of filming *Dial M for Murder.*

Martin, Pete. "Pete Martin Calls on Alfred Hitchcock." *Saturday Evening Post,* July 27, 1957. Reprinted in Harry M. Geduld, ed., *Film Makers on Film Making.* Bloomington, Ind.: Indiana University Press Paperbacks, 1969.

Roche, Catherine de la. "Conversation with Hitchcock." *Sight and Sound,* Winter 1955–56. On realism *vs.* fantasy and metaphysical elements in his work.

Torre, Marie. "Interviews." *New York Herald Tribune,* June 16, 1956.

———. "Lots of Suspense Out in the Sunshine: Mr. Hitchcock Talks of Many Things." *New York World Telegram and Sun,* February 28, 1953.

Truffaut, François, and Chabrol, Claude. "Entretien avec Alfred Hitchcock." *Cahiers du Cinéma,* no. 44 (February 1955): 19–31.

II. BOOKS

Amengual, Barthelemy, and Borde, Raymond. *Alfred Hitchcock.* Premier Plan, no. 7. Lyon: Serdoc, 1960. A brief survey.

Bogdanovich, Peter. *The Cinema of Alfred Hitchcock.* New York: The Museum of Modern Art, 1962. Interviews interspersed with commentary by Bogdanovich. Includes good stills.

Manz, Hans Peter. *Alfred Hitchcock.* Zurich: Sansouci Verlag, 1962. Mainly clips from *Cahiers* critics translated into German. Includes good stills.

Perry, George. *The Films of Alfred Hitchcock.* New York: Dutton; London: Studio Vista, 1965. Mainly a collection of superb stills.

Rohmer, Eric, and Chabrol, Claude. *Hitchcock.* Paris: Editions Universitaires, 1957. A short, pioneering, very metaphysical—and very Catholic—study.

Simsolo, Noel. *Alfred Hitchcock.* Cinema Aujourd'hui, no. 54. Paris: Editions Seghers, 1969. A brief survey with a good French bibliography.

Truffaut, François. *Le Cinéma Selon Hitchcock.* Paris: Robert Lafont, 1966. Published in English as *Hitchcock,* with the collaboration of Helen G. Scott. New York: Simon & Schuster, 1967. The famous set of interviews covering every film; interesting for the light it sheds on Truffaut as well as on Hitchcock. Still, not as definitive as its title leads one to believe.

Wood, Robin. *Hitchcock's Films.* London: A. Zwemmer; Cranbury, N.J.:

A. S. Barnes, 1965, rev. 1969; New York: Paperback Library, 1970. A pioneering reinterpretation in English, following but redefining the *Cahiers* line. Introduction and chapters on *Strangers on a Train, Rear Window, Vertigo, North by Northwest, The Birds, Marnie, Torn Curtain.* Parts of the introduction are reprinted in this volume. Also good are the chapters on *Vertigo, Strangers on a Train,* and *The Birds.*

III. ARTICLES

Agate, James. Review of *Juno and the Paycock. Around Cinemas,* 1st series. London, 1946. Reviews of *Jamaica Inn* and *Suspicion. Around Cinemas,* 2nd series. London, 1948.

Agel, Henri. *Les grands cinéastes que je propose.* Paris: Les Editions du Cerf, 1967, pp. 114–21.

Auriol, Jean-George. "*L'Ombre d'un doute* et *Lifeboat*" (reviews of *Shadow of a Doubt* and *Lifeboat*). "Festival Hitchcock," *La Revue du Cinéma* (July 1948): 64–70.

Bazin, André. "Panoramique sur Alfred Hitchcock." *Ecran Français,* no. 238 (January 23, 1950).

Bellour, Raymond. "Analyse d'un sequence de les Oiseaux." *Cahiers du Cinéma,* no. 216 (October 1969): 24–38. Detailed study of the Bodega Bay rowboat sequence with each major shot reproduced with an afterword, "Á propos," by Jean Narboni, p. 39.

———. "Ce Que Savait Hitchcock." *Cahiers du Cinéma,* no. 190 (May 1967): 32–37. Review of the Hitchcock–Truffaut interviews.

Baudrot, Sylvette. "Hitch au jour le jour." *Cahiers du Cinéma,* no. 39 (October 1954): 14–17. Journal of Hitchcock during the filming of *To Catch a Thief.*

Buchwald, Art. "Hitchcock Steps Off the Deadly Trains." *New York Herald Tribune,* January 16, 1955.

Blumenberg, Hans C. "Die Frühen Filme von Alfred Hitchcock." *Fernsehen und Film,* August–October 1969.

Bond, Kirk. "The Other Alfred Hitchcock." *Film Culture,* no. 41 (Summer 1966): 30–35. Admiration for the early silent films of Hitchcock.

Cahiers du Cinéma, no. 39 (October 1954); no. 62 (August–September 1956). Almost exclusively devoted to articles on Hitchcock. Many reprinted in *Cahiers du Cinéma in English,* no. 2 (1966).

Cameron, Ian. "Hitchcock and the Mechanics of Suspense." *Movie* 3 (October 1962): 4–7. "Hitchcock 2: Suspense and Meaning." *Movie* 6 (January 1963): 8–12. A very detailed study of the mechanics of *The Man Who Knew Too Much* (1955 version).

Chabrol, Claude. "Festival Alfred Hitchcock à la cinématheque." *Arts,* no. 571 (June 6, 1956).

———. "Histoire d'une Interview." *Cahiers du Cinéma,* no. 39 (October 1954): 39–44.

———. "Hitchcock devant le mal." *Cahiers du Cinéma,* no. 39 (October 1954): 18–24. Translated as "Hitchcock Confronts Evil." *Cahiers du*

Cinéma in English, no. 2 (1966). Important for a study of the Chabrol–Hitchcock relationship.

———. "Hitchcock aime l'Invraisemblance." *Arts,* no. 548 (December 28, 1955).

Clarens, Carlos. *An Illustrated History of the Horror Film.* New York: G. P. Putnam, 1967, pp. 167–69. Contains an interesting brief analysis of *The Birds* as a horror film.

Comolli, Jean-Louis. "Le Rideau soulevé, retombé." *Cahiers du Cinema,* no. 186 (January 1967): 36–39. Translated as "The Curtain Lifted and Fallen Again." *Cahiers du Cinéma in English,* no. 10 (May 1967): 54–55. A lively defense of *Torn Curtain.*

Corliss, Richard. *"Topaz." Film Quarterly,* Spring 1970, pp. 41–44. An interesting assessment of Hitchcock's triumphs and failures in his later films.

Demonsablon, Phillipe. "Lexique mythologique pour l'oeuvre de Hitchcock." *Cahiers du Cinéma,* no. 62 (August–September 1956): 17ff. A list of familiar Hitchcock objects and the movies in which they occur.

Domarchi, Jean. "Le chef-d'oeuvre Inconnu." *Cahiers du Cinéma,* no. 39 (October 1954): 33–38. Praise for *Under Capricorn.*

Doniol-Valcroze, Jacques. *"Rebecca," "La Maison du docteur Edwardes," "Soupçons." La Revue du Cinéma,* July 1948, pp. 72–77. Reviews of *Spellbound, Rebecca,* and *Suspicion* in the parent journal of *Cahiers du Cinéma.*

Douchet, Jean. "La Troisième clé d'Hitchcock." *Cahiers du Cinéma,* no. 99 (September, 1959): 44–50; no. 102 (December 1959): 30–37. Pioneering articles stressing suspense and magic.

———. "Hitch et son public." *Cahiers du Cinéma,* no. 113 (November 1960): 7–15. Usually considered the third in a series, and the best; deals with *Psycho* and *Rear Window.*

———. "Le Proces de Lucullus ou Hitchcock econome." *Cahiers du Cinéma,* no. 163 (February 1965): 36–45. Mainly about *Marnie.*

Durgnat, Raymond. "The Strange Case of Alfred Hitchcock." *Films and Filming* 16, no. 6 (March 1970) through vol. 17, no. 2 (November 1970). Book-length series of articles surveying Hitchcock criticism and all the films; a kind of metacriticism, often rich but diminished by a basic dislike of some of the films.

———. *"Strangers on a Train."* In "Images of the Mind." *Films and Filming,* March 1969. A provocative discussion of the endless doublings in the movie and their mysterious significance.

Dyer, Peter John. "Young and Innocent." *Sight and Sound,* Spring, 1961, pp. 80–83. Not about *Young and Innocent* but about thirties English thrillers. Good discussion of *The Man Who Knew Too Much* and *The Secret Agent.*

Godard, Jean-Luc. "Le Cinéma et son double." *Cahiers du Cinéma,* no. 72 (June 1957): 35–42. To appear in Tom Milne, ed. and trans., *Godard on Godard.* New York: Viking, forthcoming (1972). A close look at the documentary technique of *The Wrong Man.*

Godet, Sylvain. "Angoisse derrière la vitre." *Cahiers du Cinéma,* no. 186 (January 1967): 39–42. Translated as "Anxiety Behind the Window-pane" in *Cahiers du Cinéma in English,* no. 10 (May 1967): 55–56. Part of the defense of *Torn Curtain.*

Grierson, John. *Grierson on Documentary.* Edited by Forsyth Hardy. New York: Harcourt, Brace, 1947, pp. 49–52; rev. ed., Berkeley and Los Angeles: University of California Press, 1966, pp. 74–76.

Gilliat, Penelope. "The London Hitch." *The New Yorker,* September 11, 1971, pp. 91–93. Good survey of the English films of Hitchcock, stressing the value of his social realism—and noting its absence in the American films.

Gun, Joyce W. "Hitchcock et la TV." *Cahiers du Cinéma,* no. 62 (August–September 1956): 6–7.

Hardison, O. D. "The Rhetoric of Hitchcock's Thrillers." In W. R. Robinson, ed., *Man and the Movies.* Baton Rouge: Louisiana State University Press, 1967, pp. 137–52. Reprinted in a Penguin Paperback edition, 1969. An interesting article from a literary viewpoint.

Haskell, Molly. "*Stage Fright.*" *Film Comment* 6, no. 3 (Fall 1970): 49–59. An interesting view of a neglected movie. Pursues the theatrical metaphor that informs many of Hitchcock's films.

Higham, Charles. "Hitchcock's World." *Film Quarterly* 16, no. 2 (Winter 1962): 3–16. A famous attack.

Hitchcock, Alfred. "Film Production." *Encyclopedia Britannica,* vol. 15 (1958). Part 3 of "Motion Pictures." Hitchcock's most thorough discussion of film techniques and production.

———. "The Woman Who Knows Too Much." *McCall's,* March 1956.

"Hitchcock Anglais." *Cahiers du Cinéma,* no. 62 (August–September 1956). A survey of Hitchcock's English films by a variety of critics.

Houston, Penelope. "The Figure in the Carpet." *Sight and Sound* 32, no. 4 (Autumn 1963): 159–64. A major assessment of Hitchcock that finds him lacking in fullness and coherence.

Kane, Lawrence. "The Shadow World of Alfred Hitchcock." *Theatre Arts* 33, no. 4 (May 1949): 33–40. Another negative article.

Mazzocco, Robert. "It's Only a Movie." *New York Review of Books,* February 26, 1970, pp. 27–31. A review of the Truffaut book and *Topaz* that views Hitchcock as a lesser artist, primarily comic.

Millar, Gavin. "Hitchcock *versus* Truffaut." *Sight and Sound,* Spring 1969, pp. 82–87. Millar finds complexities in the suspense of Hitchcock but disputes Truffaut's claims to high seriousness and moral intentions in Hitchcock's films.

Narboni, Jean. "La machine infernale." *Cahiers du Cinéma,* no. 186 (January 1967): 35. Reprinted in *Cahiers du Cinéma in English,* no. 10 (May 1967): 51. Part of the defense of *Torn Curtain,* and very good.

Noble, Peter. "An Index to the Creative Work of Alfred Hitchcock." *Sight and Sound* supplement, Index series, no. 18 (London, 1949). Filmography with critical commentary.

Nugent, Frank S. "Mr. Hitchcock Discovers Love." *New York Times Magazine,* November 3, 1946, pp. 12–13, 63–64. A popularly oriented article discussing the addition of love themes in *Spellbound* and *Notorious* to the usual Hitchcock suspense.

Pechter, William S. "The Director Vanishes." *Moviegoer,* no. 2 (Summer–Autumn 1964). Revised with a postscript on *Turn Curtain* in *Twenty-Four Times a Second.* New York: Harper & Row, 1971, pp. 175–94. A very interesting article that attempts to explain some of the problems in Hitchcock's movies as divisions within the artist.

Perkins, V. F. *"Rope." Movie,* no. 7 (February 1963): 11–13. A fine, detailed study of the implications of Hitchcock's tracking methods in this 1940s film.

Ross, Don. "Alfred Hitchcock, a Very Crafty Fellow." *New York Herald Tribune,* March 5, 1956. On filming *The Wrong Man.*

Sarris, Andrew. Reviews of *Torn Curtain, The Birds,* and *Marnie.* In *Confessions of a Cultist.* New York: Simon & Schuster, 1971, pp. 268–72, 84–86, 141–44.

Scherer, Maurice. "Les Enchaînes" (review of *Notorious*). *La Revue du Cinéma,* no. 15 (July 1948): 70–72. Part of the "Festival Hitchcock" articles. Interesting as an early example of the *Cahiers* attitudes toward Hitchcock.

Sonbert, Warren. "Alfred Hitchcock: Master of Morality." *Film Culture,* no. 41 (Summer 1966): 35–38. On Hitchcock's high disdain for his audiences.

Stanbrook, Alan. *"The Lady Vanishes." Films and Filming,* July 1963, pp. 43–47. A look at the early reviews and a reevaluation.

Taylor, John Russell. "Hitchcock." In *Cinema Eye, Cinema Ear.* New York: Hill & Wang, 1964, pp. 170–99. A good general survey.

Techine, André. "Les naufrages de l'autocar" ("The Castaways of the Bus"). *Cahiers du Cinéma,* no. 186 (January 1967): 42. Reprinted in *Cahiers du Cinéma in English.*

Truffaut, François. "Un Trusseau de Fausses Clés." *Cahiers du Cinéma,* no. 39 (October 1954): 45–52. Reprinted in *Cahiers du Cinéma in English,* no. 2 (1966): 61–66. Truffaut's initial article on Hitchcock, written when he was twenty-one, arguing—often flamboyantly—that Hitchcock is aware of his "doubling" methods and themes, but that he lies to critics and reporters. *Shadow of a Doubt* is the chief evidence.

Walker, Michael. "The Old Age of Alfred Hitchcock." Movie, no. 18 (Winter 1971): 10–13. A very intelligent appraisal of *Topaz.*

Wood, Robin. "Psychoanalyse de *Psycho." Cahiers du Cinéma,* no. 113 (November 1960): 1–7. Earlier version of the *Psycho* chapter that appears in his book *Hitchcock's Films.*

Index